Chemistry of Foods and Beverages: Recent Developments

Academic Press Rapid Manuscript Reproduction

Papers presented at the proceedings of the Second International Flavor Conference held in Athens, Greece, July 20–24, 1981,
and
Formulated Foods and Their Ingredients: Recent Progress in Chemistry, Nutrition, and Technology held in Anaheim, California, November 1–4, 1981.

Chemistry of Foods and Beverages: Recent Developments

Edited by

George Charalambous
Agricultural and Food Chemistry Division of the American Chemical Society

George Inglett
U.S. Department of Agriculture
Science and Education Administration
Northern Regional Research Center
Peoria, Illinois

1982

ACADEMIC PRESS
A Subsidiary of Harcourt Brace Jovanovich, Publishers

New York London
Paris San Diego San Francisco São Paulo Sydney Tokyo Toronto

ACADEMIC PRESS, INC.
111 Fifth Avenue, New York, New York 10003

United Kingdom Edition published by
ACADEMIC PRESS, INC. (LONDON) LTD.
24/28 Oval Road, London NW1 7DX

Library of Congress Cataloging in Publication Data
Main entry under title:

Chemistry of foods and beverages.

1. Food--Analysis--Congresses. 2. Beverages--Analysis--Congresses. I. Charalambous, George, Date. II. Inglett, George.
TX541.C44 664 82-4043
ISBN 0-12-169080-6 AACR2

PRINTED IN THE UNITED STATES OF AMERICA

82 83 84 85 9 8 7 6 5 4 3 2 1

Contents

Contributors

Numbers in parentheses indicate the pages on which the authors' contributions begin.

Jens Adler-Nissen (149), *Enzyme Applications Research and Development, Novo Industri A/S, DK-2880 Bagsvaerd, Denmark*

Robert P. Allingham (293), *Bio-Organic Chemicals Research and Development Department, Pfizer Inc., Groton, Connecticut 06340*

Fred Austin (251), *Pembertons Food Processing Equipment, Inc., Memphis, Tennessee 38107*

Adegboyega O. Bammeke (171), *Department of Preventive and Social Medicine, University of Ibadan, Ibadan, Nigeria*

Ron Bennett (251), *Technical Service Department, Grain Processing Corporation, Muscatine, Iowa 52761*

Daniel R. Burgum (265), *Kelco Division of Merck and Co., Inc., San Diego, California 92123*

Ross C. Clark (265), *Kelco Division of Merck and Co., Inc., San Diego, California 92123*

Paul Dirinck (25), *Laboratorium voor Organische Scheikunde, Faculteit van de Landbouwwetenschappen, Rijksuniversiteit-Gent, Coupure 533, B-9000 Gent, Belgium*

Friedrich Fendesack (1), *FB 13-Lebensmitteltechnologie und Biotechnologie, Institut für Fermentation und Brauwesen, Fachgebiet Chemisch-technische Analyse, Technische Universität Berlin, D-1000 Berlin 65, West Germany*

Pio Gagliardi (49), *Carlo Erba Strumentazione, I-20090 Rodano-Milano, Italy*

L. Mark Hanover (211), *A.E. Staley Manufacturing Company, Decatur, Illinois 62525*

Judith D. Hills (101), *Department of Food Science and Technology, College of Agricultural and Environmental Sciences, University of California, Davis, Davis, California 95616*

Alan C. Hugunin (235), *Research and Development Center, Foremost-McKesson, Inc., Dublin, California 94566*

John E. Long (195), *Archer Daniels Midland Company, Cedar Rapids, Iowa 52406*
Mitsuru Motono (181), *Food Product Development Laboratories, Central Research Laboratories, Ajinomoto Co., Inc., Kawasaki, 210 Japan*
Hans Sejr Olsen (149), *Enzyme Applications Research and Development, Novo Industri A/S, DK-2880 Bagsvaerd, Denmark*
Angela Reinke (1), *FB 13-Lebensmitteltechnologie und Biotechnologie, Institut für Fermentation und Brauwesen, Fachgebiet Chemisch-technische Analyse, Technische Universität Berlin, D-1000 Berlin 65, West Germany*
Gerald F. Russell (101, 129), *Department of Food Science and Technology, College of Agricultural and Environmental Sciences, University of California, Davis, Davis, California 95616*
Eugene Sander (251), *Technical Service Department, Grain Processing Corporation, Muscatine, Iowa 52761*
Niceas Schamp (25), *Laboratorium voor Organische Scheikunde, Faculteit van de Landbouwwetenschappen, Rijksuniversiteit–Gent, Coupure 533, B-9000 Gent, Belgium*
Takayuki Shibamoto (73), *Department of Environmental Toxicology, College of Agricultural and Environmental Sciences, University of California, Davis, Davis, California 95616*
Robert E. Smiles (305), *Technical Service Center, Chemicals Division, Pfizer Inc., Groton, Connecticut 06340*
Oak B. Smith (323), *Wenger International, Inc., Kansas City, Missouri 64108*
M. K. C. Sridhar (171), *Department of Preventive and Social Medicine, University of Ibadan, Ibadan, Nigeria*
Gail D. Teague (265), *Kelco Division of Merck and Co., Inc., San Diego, California 92123*
Roland Tressl (1), *FB 13-Lebensmitteltechnologie und Biotechnologie, Institut für Fermentation und Brauwesen, Fachgebiet Chemisch-technische Analyse, Technische Universität Berlin, D-1000 Berlin 65, West Germany*
Giuseppe Romano Verga (49), *Carlo Erba Strumentazione, I-20090 Rodano-Milano, Italy*
L. Steven Young (195), *Archer Daniels Midland Company, Decatur, Illinois 62525*

Preface

The chemistry of foods and beverages is of continual importance to growers, processors, manufacturers, and, ultimately, the consumer. The chemistry of foods and beverages is important in understanding most aspects of foods and beverages including their ultimate fate—consumer acceptability. Regardless of the other numerous desirable attributes of a successfully marketed food or beverage, an acceptable flavor is still a major asset. Other food constituents and ingredients are also important in the formulation of foods and beverages. Rapid changes in food chemistry and technology require frequent updating of recent progress. This book is a compilation of selected papers from two conferences. The first conference was concerned with the quality of foods and beverages which was the theme of the Second International Flavor Conference held in Athens, Greece, July 20–24, 1981. The conference, held under the auspices of the Hellenic Republic Ministry of Agriculture, was cosponsored by the Agricultural and Food Chemistry Division of the American Chemical Society, the Institute of Food Technologists, and the American Society of Flavor Chemists, Inc. It was cohosted by the Food Chemistry Department of the University of Ioannina, the Cereal Institute, Thessaloniki, and the Greek Institute of Food Scientists.

The second conference, Formulated Foods and Their Ingredients: Recent Progress in Chemistry, Nutrition, and Technology, was concerned with recent progress in the chemistry and technology of formulated foods and their ingredients and was held in Anaheim, California, November 1-4, 1981. The conference was sponsored by the Division of Agricultural and Food Chemistry of the American Chemical Society, the Institute of Food Technologists, the American Association of Cereal Chemists, and the International Association for Cereal Chemistry.

The topics of this book cover these wide-ranging subjects: aroma components of hops contribution to beer flavor; headspace analysis for the evaluation of fresh fruits; effect of fast indigenization on the quality of foods and beverages; headspace analysis of flavors with capillary column and multidetector systems; bitterness and other flavor qualities of protein hydrolysates; computer-assisted quantitation of carrot volatiles; a micro-olfactometer for chemical sensory analysis; the use of fused silica capillary columns for flavor analysis; polydextrose update; soy protein isolates uses based on unique processing and formulation techniques; the usages of flavor nucleotides in foods; 90% high fructose corn syrup as a food ingredient for the diabetic; dairy-based ingredients as alternatives to traditional sweeteners; functionality of corn-derived sweeteners; applications of xanthan gum in food systems; and criteria for the selection of ingredients for use in extrusion-cooked formulated foods.

This volume should be a useful book for students, chemists, technologists, and executives who are involved with any facet of foods and beverages.

We wish to thank all who attended and participated in the conferences, and particularly the speakers who contributed papers. The publisher's guidance and assistance are also greatly appreciated.

CHANGES OF AROMA COMPONENTS DURING STORAGE AND PROCESSING OF HOPS AND THEIR CONTRIBUTION TO BEER FLAVOR

Roland Tressl
Friedrich Fendesack
Angela Reinke

Technische Universität Berlin

I. INTRODUCTION

Even today, despite advances in brewing technology, the production of hop-aromatic beers remains difficult. Most of the brewing technologists are of the opinion, that hop oil components are not transferred to the beer. In 1978 we characterized more than 100 hop aroma components in the aroma extract of a German Pilsner beer which possessed a desirable hoppy aroma (1). Hop aroma components are transferred to beer according to their hydrophylic/lipophylic characters. Therefore, only oxygenated components were determined in beer. During hop storage and processing, hydrocarbons are transformed to some extent into oxygenated components and carotenoids are split into contributing norcarotenoids. Therefore, the production of hop-aromatic beers is influenced by the hop variety and its storage and processing condition as well as by the yeast strain and the brewing technology used.

Present address: Technische Universität Berlin
Seestrasse 13, D-1000 Berlin 65

ISBN 0-12-169080-6

II. INVESTIGATION OF HOP AROMA COMPONENTS IN PILSNER BEER

When we began our hop research project, we found few analytical reference points on the extent to which hop aroma constituents are transferred to beer. The available techniques could not prove that the major sesquiterpenes were present in beer. This led to the assumption that sensorially relevant oxygenated hop oil constituents were not present in beer either. Buttery and Black (2) identified some hop derived esters and ketones in an American beer. However, Sandra (3) found only few degradation products of humulones in a Belgian lager beer. We studied a Bavarian Pilsner (1) that possessed a desirable hoppy aroma and characterized more than one hundred hop derived constituents. Investigation of hop aroma constituents in beer is difficult because the trace constituents are masked by the high concentrations of yeast metabolites and Maillard reaction products. However, LSC separation according to functional groups and preparative GC were used to effectively characterize hop aroma constituents in beer.

Figure 1 shows the methods which have been used for isolation, separation, and identification of hop aroma constituents in beer.

Figure 2 presents a capillary gas chromatographic separation of the LSC fraction IV. It can be seen that more than 130 constituents (among them: hop ether (30), karahana ether (31), linalool (49), ß-fenchyl alcohol (53), α-terpineol (66), humulene epoxide I (104), humulol (120), humulenol II (132) were characterized. In addition to those found in LSC fraction IV hop aroma constituents were also detected in fractions V and VI. Some of the results and methods used have been published in detail (1). The concentration of hop aroma constituents in Pilsner beers varied considerably. In Table I the results for two German Pilsner beers are summarized. It can be seen that only oxygenated hop oil constituents are transferred to beer. The strongly concentrated hydrocarbons myrcene, caryophyllene, and humulene were not detected in the beer aroma extracts.

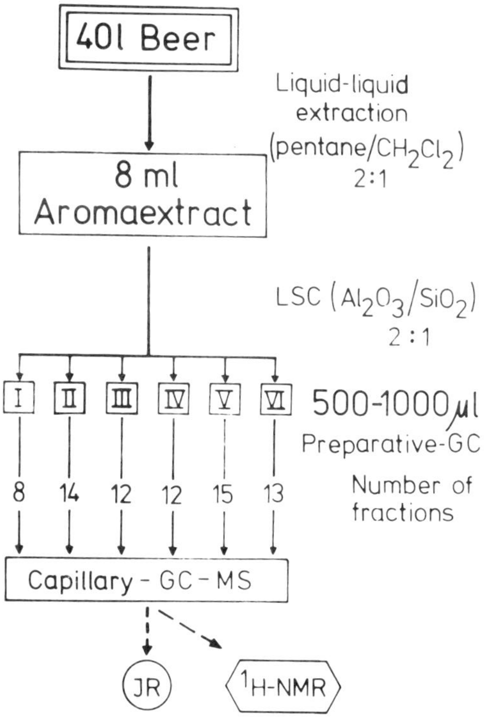

Figure 1 Characterization of hop aroma components in beer

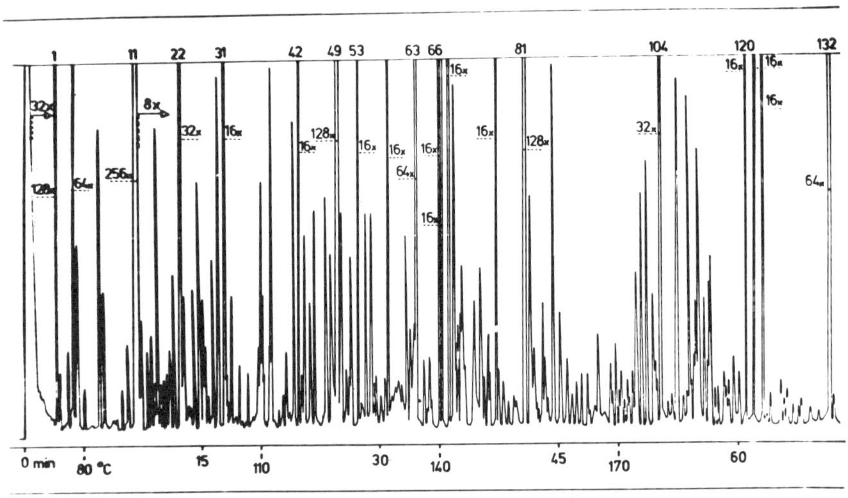

Figure 2 Gas chromatogram of LSC fraction IV, column CW20M, 50 m (0,25 mm i.d.)

hop ether (30), karahana ether (31), linalool (49), ß-fenchylalcohol (53), α-terpineol (66), humulene epoxide I (104), humulol (120), humulenol II (132)

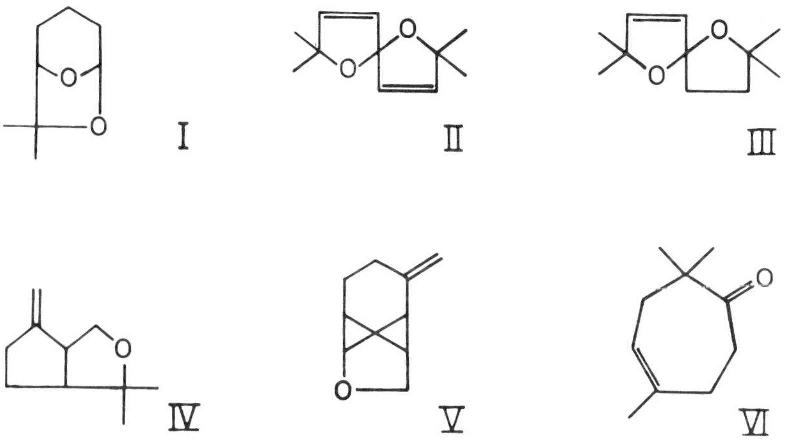

Figure 3 Cyclic ethers identified in beer. I, 7,7-Dimethyl-6,8-dioxabicyclo-(3.2.1)octane; II, 2,2,7,7-tetramethyl-1,6-dioxaspiro(4,4)nona-3,8-diene; III, 2,2,7,7-tetramethyl-1,6-dioxaspiro-(4,4)nona-3-ene; IV, hop ether; V, karahana ether; VI, karahananaenone

Table I Hop Aroma Components Characterized in Pilsner Beer (ppb)

Component	I_K (CW20M)	M	Beer A	Beer B
1 linalooloxide	1462	152	+	20
2 hop ether	1323	152	+	35
3 karahana ether	1333	152	10	60
4 campher	1484	152	+	+
5 7,7-dimethyl-6,8-dioxa-bicyclo(3.2.1)octane	1261	142	10	50
6 2,2,7,7-tetramethyl-1,6-dioxaspiro(4.4)-nona-3,8-diene	1292	180	10	10
7 cis-jasmone	1908	164	+	20
8 caryophyllene epoxide	1938	220	10	18
9 humulene epoxide I	1963	220	20	125
10 humulene epoxide II	1984	220	60	40
11 humuladienone		220	+	10
12 linalool	1494	154	80	470
13 terpinen-4-ol	1558	154	20	15
14 myrcenol		154	+	-
15 β-fenchol	1526	154	+	40
16 α-terpineol	1641		60	40
17 citronellol	1710	156	30	10
18 junenol	2005	222	10	5
19 caryolan-1-ol	1986	222	10	25
20 nerolidol	1990	222	20	25
21 epicubenol	2028	220	10	20
22 caryophyllenol		220	+	5
23 humulol	2098	222	50	220
24 δ-cadinol	2138	222	+	35
25 T-cadinol	2170	222	20	45
26 humulenol II	2218	222	20	1150
27 citronellyl acetate	1610	198	20	-

Table I

	I_K (CW20M)	M	Beer A	Beer B
28 neryl acetate	1686	196	10	-
29 ethyl dec-4-enoate	1655	198	100	35
30 3-methylbutyl dec-4-enoate	1881	240	10	20
31 ethyl deca-4,8-dienoate	1696	196	20	15
32 4,4-dimethyl-2-buten-4-olide	1558	112	500	1750
33 4,4-dimethyl butan-4-olide	1528	114	100	160

Table II Changes of Components during Fermentation

Component (50 ppm)	I_K	Medium	Yeast
1 trans-2-hexenoic acid	1346	46,1	-
2 3-decenoic acid	1627	10	12
3 2-heptanone	1165	20,3	1
4 2-undecanone	1578	2	16,5
5 pulegone	1614	24,8	20
6 caryophyllene epoxide	1948	5,4	40
Product			
1' ethyl hex-2-enoate	1336	1	-
2' ethyl dec-3-enoate	1617	1,2	4
3' heptanol-2	1275	17,2	6,5
4' undecanol-2	1672	4,7	22,5

Beer A was a German Pilsner brewed with a mixture of Spalter and Northern Brewer hops. Beer B was brewed in the Hallertau with Hallertauer Mittelfrüh hops. In our opinion the hop aroma constituents in beer are influenced by the variety and the storage conditions of the hops or hop products, as well as by the yeast and brewing technology. Hop ether, karahana ether, caryophyllene and humulene epoxides, linalool, humulol and humulenol are strongly concentrated components in Pilsner beer B. They are responsible for the hoppy aroma of the beer A. The threshold values of these components (in beer) were determined for linalool at 80 ppb and for humulenol II at 500 ppb. According to Peacock at al. (4) humulenol II contributes to the hoppy aroma of beer.

Figure 3 presents the chemical structures of cyclic ethers which have been characterized in certain hops and Pilsner beers. The concentrations of cyclic ethers and of sesquiterpene epoxides increased considerably during the storage of hops (5).

In Figure 4 humulene derived components are summarized. The concentrations of the oxygenated constituents increased during the storage of hops. Humuladienone (III), humulene epoxide I (V), humulene epoxide II (VI), humulenol I (VIII), humulenol II (IX), and humulol (VII) were identified in the aroma extracts of Pilsner beers. During the storage of hops caryophyllene is transformed to some extent into caryophyllene epoxide, caryolan-1-ol, and caryophyllenol which were characterized in Pilsner beers. As shown in Figure 5 hop oil constituents are transferred to beer according to their lipophylic/hydrophylic characters. Hydrocarbons, esters, and ketones with higher carbon chains are adsorbed by the yeast.

Lactones, alcohols, ethers, and epoxides are transferred to some extent to beer. In addition to the transesterfication of methyl esters into ethyl and isoamylesters, we observed the reduction of ketones to secondary alcohols and the hydrogenation of C = C bonds and epoxides. This has been demonstrated in model systems. Therefore, only the oxygenated constituents of the hop oil are contributors to the hoppy aroma of beer. Table II illustrates some results for the esterfication of fatty

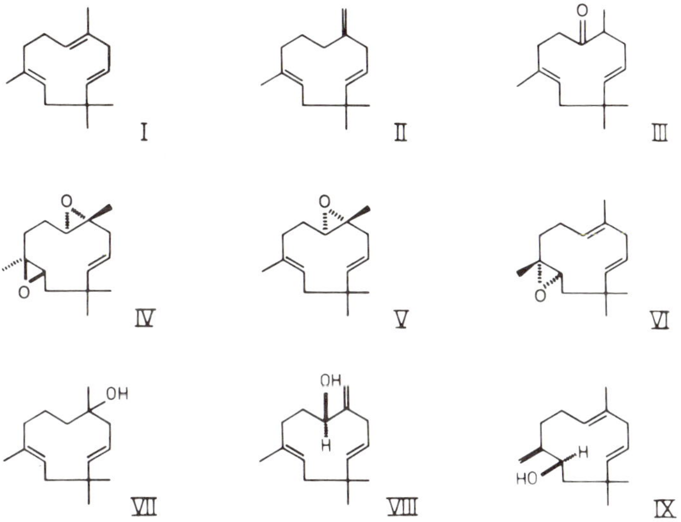

Figure 4 Sesquiterpenoides identified in beer. I, α-humulene; II, ß-humulene; III, humuladienone; IV, humulene dioxide; V, humulene epoxide I; VI, humulene epoxide II; VII, humulol; VIII, humulenol I; IX, humulenol II.

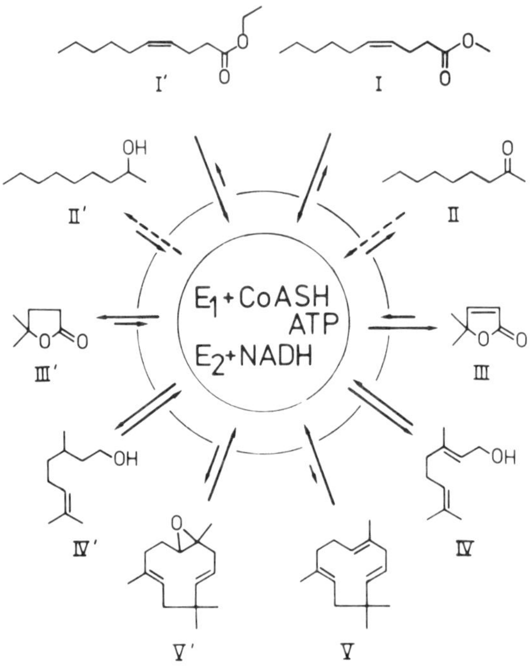

Figure 5 Most probable pathways for changes of hops aroma constituents by yeast metabolism. Transesterfication, reduction of carbonyls, hydrogenation of C=C bonds, and distribution of components according to their lipophylic/hydrophylic characters

acids and the reduction of ketones during fermentation as well as the distribution of components between medium and yeast. It can be seen that 2-hexenoic acid and the ethyl ester are completely transferred to beer. In contrast, 55 % of the 3-decenoic acid and 76 % of the corresponding ethyl dec-3-enoate were adsorbed by the yeast. Similar results were obtained with 2-heptanone and 2-undecanone and the corresponding alcohols respectively. Pulegone and caryophyllene epoxide were not reduced by this yeast strain.

III. CHANGES OF AROMA COMPONENTS DURING STORAGE AND PROCESSING OF HOPS

Table III presents some results for oxygenated hop oil constituents of four aroma hops and one bitter hop. It can be seen that the concentrations of hop ether, karahana ether, caryophyllene and humulene epoxides differ considerably. The varieties Tettnanger Aroma and Hersbrucker Spät contain low amounts of humulene epoxides. Bitter hops like Golden Brewer possess high amounts of cyclic ethers and epoxides but also high concentrations of neral and geranial which may be responsible for the formation of certain off-flavors.

The five hop varieties were stored under similar conditions and investigated by capillary GC-MS. The results and methods used will be published in detail (6). It can be seen from Table III that the contents of terpene and sesquiterpene alcohols differ considerably. The variety Hersbrucker Spät contains low amounts of humulenol but high amounts of globulol, guaiol, α-eudesmol, and ß-eudesmol which were not detected in the other varieties. Hallertauer Mittelfrüh possessed the highest concentrations of humulenol which has been characterized as a hoppy aroma component in beer.

The changes of some hop oil components during processing and storage of hops are shown in Table IV. Hallertauer Mittelfrüh hops were harvested and stored at 0°C in a dark chamber. One part of the hops was processed to pellets and stored at 0° - 20°C. During a storage period of one year, the hydrocarbon level of natural hops decreased consi-

Table III Oxygenated Components in Different Hop Varieties

Component	A	B	C	D	E
esters, ketones					
1 2-methylpropyl 2-methylpropionate	680	190	150	50	3350
2 methyl 4-methyl-hex-2-enoate	660	190	590	760	300
3 methyl dec-4-enoate	600	320	450	730	830
4 methyl deca-4,8-dienoate	130	50	140	220	650
5 methyl geranate			250		880
6 4-methylpentan-2-one	120	70	60	90	60
7 3-methylpentan-2-one	40	50	150	160	80
8 2-nonanone	480	170	320	410	60
9 2-decanone	290	150	170	400	50
10 2-undecanone	700		520	630	270
ethers, epoxides, carbonyls					
11 6-methyl-5-hepten-2-one	28	27	32	120	80
12 "Oxetone"	25	10	20	40	90
13 hop ether	25	30	30	50	95
14 karahana ether	20	10	30	50	85
15 geranial	20	3	20	45	160
16 neral	30	10	17	50	110
17 ß-ionone	4	3	5	5	3
18 humuladienone	85	35	80	130	30
19 caryophyllene epoxide	140	120	110	45	210
20 humulene epoxide I	80	70	50	25	95
21 humulene epoxide II	340	330	130	140	390
terpene, sesquiterpene alcohols					
22 linalool	730	290	420	530	580

Table III

Component	A	B	C	D	E
23 α-terpineol	100	50	50	85	95
24 geraniol	105	50	60	160	940
25 caryolan-1-ol	85	40	35	35	110
26 epicubenol	50			25	110
27 globulol	-	-	180	-	-
28 humulol	175	120	85	150	155
29 δ-cadinol	155	120	-	25	100
30 T-cadinol	50	85	-	50	50
31 guaiol	-	-	140	-	-
32 α-cadinol	160	100	-	100	120
33 α-eudesmol	-	40	340	-	-
34 ß-eudesmol	-	80	170	-	-
35 humulenol	110	310	30	85	120
36 farnesol	-	-	-	90	-

A = Spalt

B = Hallertauer Mittelfrüh

C = Hersbrucker Spät

D = Tettnanger Aroma

E = Golden Brewer

Table IV Changes in Concentration[a] of Aroma Constituents during Processing and Storage of Hops[b]

	Form and Years						
	Natural		Pellets				Pressed
	1977	1978	1977	1978A[c]	1978B[d]	1978C[e]	1978D[f]
hop oil (mg/100 g)	800	570	880	680	620	500	550
hydrocarbons (%)	86	68	81	88	82	73	67
oxygenated components (%)	14	32	19	12	18	27	33
myrcene	2,600	810	2,050	2,100	860	840	400
caryophyllene	1,250	430	1,110	750	490	510	490
humulene	2,370	1,000	2,050	1,530	1,080	1,160	1,090
caryophyllene epoxide	45	60	42	18	11	70	110
humulene epoxide I	20	55	26	30	45	50	70
humulene epoxide II	160	220	165	65	42	230	370
linalool	35	17	36	22	16	18	17
humulol	40	25	50	25	28	26	25
humulenol	50	110	60	30	6	50	70
karahana ether	-	5	-	3	3,5	3,5	5

a In milligrams per kilogram unless otherwise indicated.
b Variety Hallertauer Mittelfrüh.
c One-year storage, 0°C, inert atmosphere.
d One-year storage, 20°C, inert atmosphere.
e One-year storage, 0°C, low level of atmospheric oxygen.
f Hop powder, one-year storage, 0°C, low level of atmospheric oxygen.

derably and the epoxides and alcohols increased significantly. Caryophyllene and humulene were transformed into the corresponding oxygenated components. During the processing of pellets, in which the hops are macerated, pressed, heated and evaporated, the hop oil content remained nearly constant. After a storage period of one year at 0°C without oxygen (Pellets 1978A), the hop oil content per 100 g had decreased from 0,88 g to 0,68 g. The ratio of hydrocarbons to oxygenated comp. remained const.. Therefore, the hop aroma constituents had decreased by polymerization. The sensorially relevant oxygenated constituents decreased considerably. This trend is more clearly observed at a storage temperature of 20°C (Pellets 1978B).

After storage for one year at 0°C with a small amount of oxygen (Pellets 1978C) the hop oil content had decreased to 0,5 g/100 g. Hydrocarbons had decreased significantly and oxygenated components had increased considerably. The important constituents caryophyllene epoxide and humulene epoxide increased 8-fold to 10-fold compared to their level in Pellets 1978B. The last column shows results for a hop powder stored at 0°C with a small amount of oxygen. The larger surface area of these samples favours the transformation of undesirable hydrocarbons into sensorially important hop aroma constituents.

These experiments show that the composition of hop aroma components changes during storage and processing. Surface area, oxygen content, temperature, storage time, and light influence the transformation of hydrocarbons into oxygenated components.

In Germany hop pellets and methylenchloride extracts each account for about one third of the total hop production. In the near future methylene chloride extraction which has been used for the extraction of bitter hops for many years may be substituted by other solvents. Carbon dioxide in the hyper-critical stage or ethanol may be applied in hop extraction. We therefore investigated aroma components in hop extracts which had been produced from Hersbrucker Spät hop pellets. Some of the results are shown in Table V. The CO_2-extract contained similar hop oil components to pellets. It can be seen that the concentrations of myrcene and

Table V Aroma Components in Hop Extracts

		CO_2-E	CH_2CL_2-E	Ethanol-E
Hop oil content mL/100 g hop		0,5 mL	0,48 mL	0,5 mL
Component				
1	myrcene	2480	47	7
2	α -copaene	42	20	22
3	ß-caryophyllene	910	300	330
4	aromadendrene	180	85	100
5	humulene	1850	800	880
6	ß-selinene	810	150	180
7	α -selinene	350	150	180
8	γ -cadinene	240	97	115
9	selina-4,7-diene	620	230	240
10	selina-3,7-diene	150	15	15
11	methyl 4-methyl-hex-2-enoate	42	12	12
12	methyl nonanoate	40	26	30
13	methyl dec-4-enoate	250	105	135
14	methyl deca-4,8-dienoate	130	35	40
15	methyl dodecadienoate	50	12	14
16	2-nonanone	45	30	15
17	2-undecanone	290	210	250
18	2-tridecanone	185	125	135
19	2-tetradecanone	32	25	25
20	2-pentadecanone	200	220	300
21	hop ether	5	8	8
22	karahana ether	3	6	6
23	ß-ionone	32	18	30
24	caryophyllene epoxide	100	100	110
25	humulene epoxide I	65	45	80
26	humulene epoxide II	85	70	75

Table V (continuned)

27	2-methyl-3-buten-2-ol	65	90	185
28	prenylalcohol	10	65	90
29	linalool	135	90	110
30	globulol	16	8	10
31	guaiol	40	35	50
32	γ -cadinol	55	45	50
33	α -eudesmol	140	100	115
34	ß-eudesmol	120	90	115

sesquiterpene hydrocarbons decreased considerably in the CH_2Cl_2 and ethanol extracts. Methyl esters showed the highest concentrations in the CO_2-extract and decreased significantly in the ethanol and CH_2Cl_2-extracts. Ketones and cyclic ethers as well as sesquiterpene epoxides were present in similar amounts. Linalool and sesquiterpene alcohols - which are transferred to beer - showed the highest concentrations in the CO_2-extract. Prenylalcohol and 2-methyl-3-buten-2-ol which are known as degradation products of humulone increased in the ethanol extract. These results indicate that CO_2-extracts may have some advantages for the productions of hop-aromatic beers in the future.

IV. NORCAROTENOIDS IN HOP AND BEER

During our storage experiments we characterized some norcarotenoids which are known as important aroma constituents in Bulgarian rose oil (7), tea (8), raspberries (9), tobacco (10), and passion fruit (11). We therefore investigated the formation of these constituents in hops and tried to identify some of them in beer. Figure 6 illustrates some norcarotenoids that have been characterized in stored hops and hop-aromatic beers. 6-Methyl-5-hepten-2-one, ß-ionone, and 5,6-epoxy-ß-ionone were characterized in stored hops at 50, 5 and 10 mg/kg, respectively. On the other hand components IV - VI were found in trace amounts. 3-Hydroxy-ß-damascone is known as a precursor of ß-damascenone. In Pilsner beer, the amounts present were smaller. The hydrophylic 3-hydroxy-ß-damascone was determined at 100 µg/kg and ß-damascenone at 20 - 30 µg/kg. The concentrations of ß-ionone, 5,6-epoxy-ß-ionone and dihydroactindiolide were 2 - 3, 35, and 35 µg/kg, respectively. ß-Ionone, 5,6-epoxy-ß-ionone and dihydroactindiolide are formed from ß-carotene (12). Isoe et al. (12) demonstrated the photo-oxidative splitting of ß-carotene into ß-ionone and dihydroactindiolide. They showed that triplet oxygen reacts with ß-carotene to form ß-ionone. This reaction was carried out in the dark without sensitizer. Lipid oxidation during storage of hops may catalize this reaction. The formation of dihydroactindiolide from ß-carotene or ß-ionone needs singlet oxygen (irradiation and sensitizer like chlorophyll).

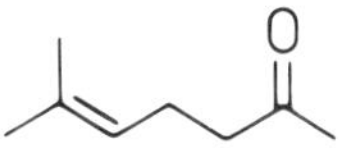

I

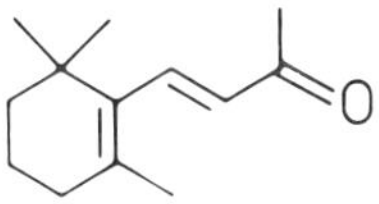

II

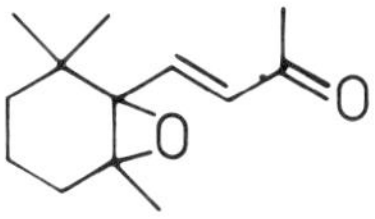

III

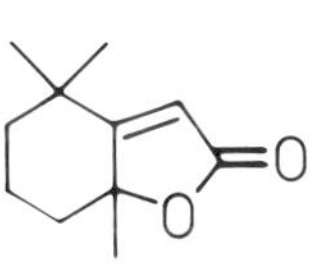

IV

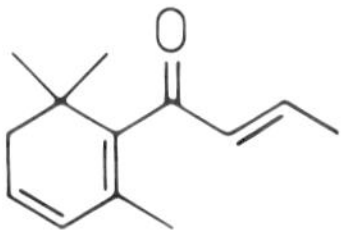

V

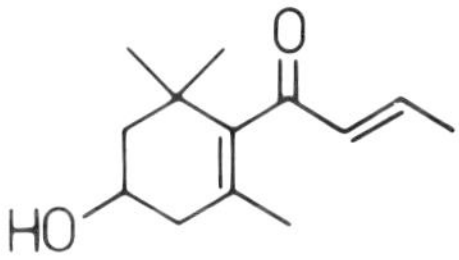

VI

Figure 6 Norcarotenoids identified in stored hops and in hop - aromatic beer.
I, 6-Methyl-5-hepten-2-one; II, ß-ionone; III, 5,6-epoxy-ß-ionone; IV, dihydroactinidiolide; V, ß-damascenone; VI, 3-hydroxy-ß-damascone

Figure 7 shows some norterpenoids formed from ß-carotene in model systems. 2,2,6-Trimethyl-5-hexenone, ß-cyclocitral, ß-ionone, and dihydroactindiolide were identified for the first time in stored hops. Figure 8 shows a reaction scheme which may explain the formation of 3-hydroxy-ß-damascone and ß-damascenone. Oxidative splitting of neoxanthine is thought to be involved in the formation of Grasshopper ketone which is easily transformed via allenic triol into 3-hydroxy-ß-damascone and ß-damascenone (13). However, ß-ionone and 3-hydroxy-ß-ionone may also act as precursors for 3-hydroxy-ß-damascone and ß-damascenone (7). The characterized components possess low thresholds and are contributing components in hops and hop aromatic beers.

V. CHANGES OF HOP AROMA COMPONENTS DURING AGING OF BEER

At present the production of Pilsner beers possessing a desirable hoppy aroma is still difficult. On the other hand, hop aromatic beers loose their typical aroma during storage and aging. Therefore, we investigated the fate of some hop aroma constituents during aging of a Pilsner beer, which had a strong hoppy aroma. Some of the results are summarized in Table VI. The beer was stored for 8 months at 4°C. During this aging lipid oxidation products and Maillard reaction products increased considerably. It can be seen from Table VI, that the concentrations of cyclic ethers and epoxides decreased slightly and that the amounts of ß-ionone and ß-damascenone increased during aging. The concentrations of linalool and of sesquiterpene alcohols, especially of humulol and humulenol decreased considerably. Therefore, these changes may contribute to the loss of hoppy aroma during aging of Pilsner beers.

ACKNOWLEDGMENTS

This work was supported by ARBEITSGEMEINSCHAFT INDUSTRIELLER FORSCHUNGSVEREINIGUNGEN, Köln.

Table VI Changes of Hop Aroma Constituents during Aging of Beer

	Component	I_K	M	Beer A (ppb)	Beer B (ppb)
1	linalooloxide	1462	204	20	-
2	hop ether	1323	152	35	30
3	karahana ether	1333	152	60	40
4	campher	1484	152	+	-
5	7,7-dimethyl-6,8-dioxa-bicyclo(3.2.1)octane	1261	142	50	20
6	2,2,7,7-tetramethyl-1,6-dioxaspiro(4.4)-nona-3,8-diene	1292	180	10	-
7	cis-jasmone	1908	164	20	-
8	caryophyllene epoxide	1938	220	18	-
9	humulene epoxide I	1963	220	125	110
10	humulene epoxide II	1984	220	40	30
11	humuladienone		220	10	-
12	linalool	1494	154	470	10
13	terpinen-4-ol	1558	154	15	20
14	myrcenol		154	-	-
15	ß-fenchol	1526	154	40	60
16	α-terpineol	1641		40	90
17	citronellol	1710	156	10	+
18	junenol	2005	222	5	-
19	caryolan-1-ol	1986	222	25	-
20	nerolidol	1990	222	25	-
21	epicubenol	2028	220	20	20
22	caryophyllenol		220	5	-
23	humulol	2098	222	220	-
24	γ-cadinol	2138	222	35	-
25	T-cadinol	2170	222	45	-
26	humulenol II	2218	222	1150	250
27	ß-ionone			2	3
28	ß-damascenone			20	30

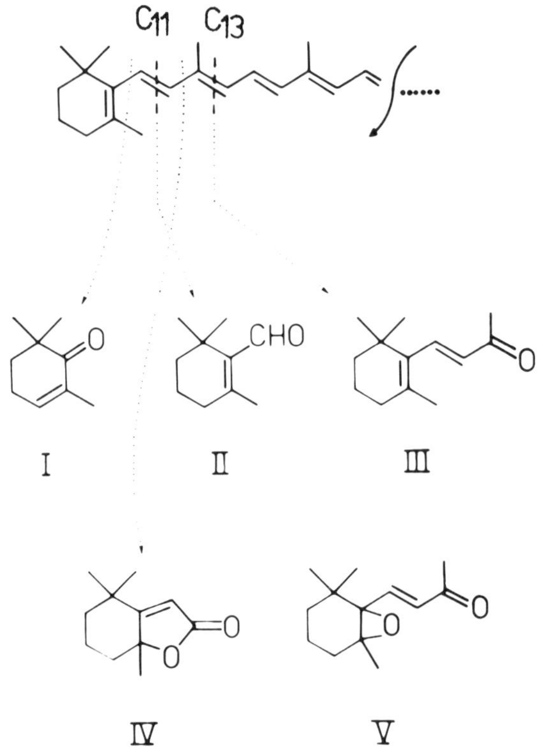

Figure 7 Formation of norcarotenoids by oxidative cleavage of ß-carotene.
I, 2,2,6-Trimethyl-5-hexenone; II, ß-cyclocitral; III, ß-ionone, IV, dihydroactindiolide; V, 5,6-epoxy-ß-ionone

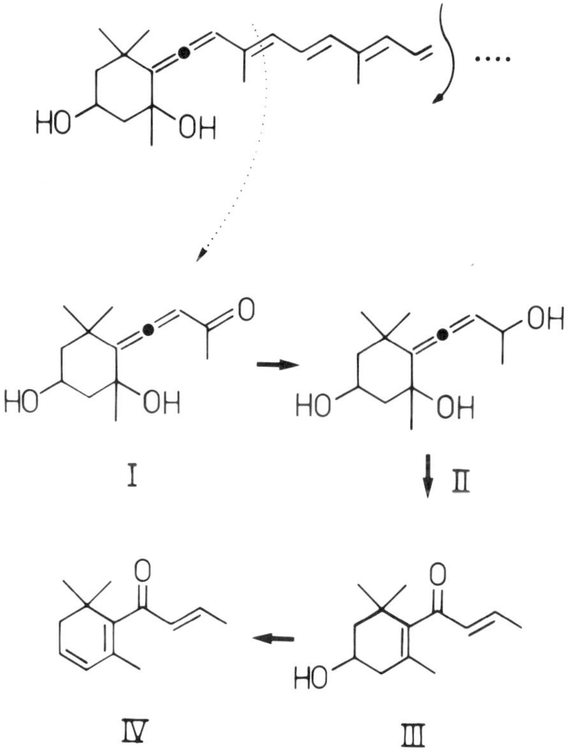

Figure 8 Reaction scheme that may explain the formation of 3-hydroxy-ß-damascone and ß-damascenone from neoxanthin.

REFERENCES

1. Tressl, R., Friese, L., Fendesack, F., Köppler, H., J. Agric. Food Chem. 26, 1422 (1978)

2. Buttery, R.G., Black, D.R., J. Food Sci. 32, 414 (1967)

3. Sandra, P., Cerevisiae 1, 37 (1976)

4. Peacock, V.E., Deinzer, M.L., McGill, L.A., Wrolstad, R.E., J. Agric. Food Chem. 28, 774 (1980)

5. Tressl, R., Friese, L., Fendesack, F., Köppler, H., J. Agric. Food Chem. 26, 1426 (1978)

6. Tressl, R., Fendesack, F., in preparation

7. Ohloff, G., Perfumer and Flavorist 3, 11 (1978)

8. Renold, W., Näf-Müller, R., Keller, U., Willhalm, B., Helv. Chim. Acta 57, 1301 (1974)

9. Winter, M., Enggist, P., Helv. Chim. Acta 54, 1891 (1971)

10. Fujimori, T., Kasuga, R., Matsushita, H., Kaneko, H., Noguchi, M., Agric. Biol. Chem. 40, 303 (1976)

11. Winter, M., Klöti, R., Helv. Chim. Acta 55, 1916 (1972)

12. Isoe, S., Hyeon, S.B., Sakan, T., Tetrahedron Lett. 1969, p. 279

13. Isoe, S., Katsumura, S., Sakau, T., Helv. Chim. Acta 56, 1514 (1974)

THE USE OF HEADSPACE CONCENTRATION ON TENAX FOR OBJECTIVE FLAVOR QUALITY EVALUATION OF FRESH FRUITS (STRAWBERRY AND APPLE)

Niceas Schamp
Paul Dirinck

Laboratory of Organic Chemistry
Faculty of Agricultural Sciences
State University of Gent
Gent, Belgium

ABSTRACT

Objective evaluation of food flavor quality by means of instrumental analyses is one of the major challenges for the flavor chemist. The fact that up to now realistic results have been obtained only for a few foodstuffs is due to the difficulty of the problem. Important aspects are: (1) isolation of a representative aroma mixture; and (2) determination of the flavor importance of the different compounds. In order to estimate the importance of the different flavor components for the overall flavor the safest procedure is to perform a relatively large number of comparisons between sensory analyses (taste testing) and chemical analyses (GC-MS analyses). For flavor quality evaluation of fresh fruits a headspace concentration technique, which isolates the volatiles released when macerating the fruits, is much more meaningful than total volatile analysis resulting from distillation or extraction procedures. The complete method consists of a combination of headspace enrichment on Tenax, high resolution gas chromatography, mass spectrometric identification, quantification on a FID-detector and specific analyses of sulphur containing compounds.

ISBN 0-12-169080-6

For different comparisons of strawberry varieties the sensory flavor intensity data could be related to the sum of the concentration of the volatiles released when macerating the fruits or, especially when "older varieties" were included, to the amount of sulphur containing compounds. For Golden Delicious apples attention has been focused on the evolution of flavor quality as a function of storage time.

INTRODUCTION

In the field of fruit quality a lot of attention has been paid to external quality attributes such as size, shape, color, and the absence of external defects. The major aim in developing new varieties and cultivars has been improvement of production, disease resistance, and external quality and very little attention has been given to internal quality attributes, such as flavor and consistency. The result of this policy is that in many cases external quality has reached a very satisfactory level, in contrast to the internal organoleptic quality, which sometimes is at the limit of acceptability. As the main reason for neglecting the internal sensory quality is its subjective character, the first step in solving this unhappy situation is the development of methods to measure flavor quality on an objective base. The purpose of this work is to demonstrate the possibilities of the described methods for objective measurement of flavor quality of different strawberry varieties and of Golden Delicious apples.

The organoleptic quality of a food is a combination of an impression on the tongue (taste), in fruits mainly determined by the acid-sugar ratio and an impression in the nose (aroma), due to the volatile organic compounds. Furthermore, flavor quality is also influenced by mouthfeel sensations. Since in the case of fruits aroma quality constitutes a major portion of the overall sensory quality in this study we concentrate on the volatile organic compounds and the use of gas chromatographic data for predicting sensory quality.

Relating sensory evaluations to physico-chemical analyses is not a simple task. First, quantitative minor components may be of critical importance to the foodstuff flavor. The very important variation (1-100 million) of our sense of odor to organic compounds is well known. Threshold values of 0.002 ppb, as in the case of 2-methoxy-3-isobutylpyrazine, remeniscent of green peppers, may be encountered [1]. A second major problem is establishing which component or components out of a complex mixture are responsible for flavor quality or a particular flavor character. The well-known approach, in which the effluent of a nondestructive detector or a flame ionization detector with splitter is sniffed, may be helpful but suffers from some important shortcomings. Flavor is mostly an integrated response to a complex mixture of volatile and nonvolatile constituents. By sniffing the effluent one deals with a differential response of separated compounds. In our opinion the safest procedure to estimate the importance of different components for the overall flavor is to establish a relatively large number of comparisons between sensory analyses (taste testing) and chemical analyses (e.g., GC-MS analyses). A third major problem is obtaining a flavor mixture which reflects the composition of volatiles that we unconsciously send to the nose while eating our foodstuffs. Before injection into the gas chromatograph enrichment of volatiles and separation from water and nonvolatile material is necessary. For objective flavor quality measurement isolation of a representative flavor mixture is of the highest importance. Clasiscal procedures, such as steam distillation and extraction, cannot be used as they cause qualitative and quantitative changes in volatile composition and do not take into account the relative volatility of the flavor substances. Indeed, flavor perception is primarily dependent upon the amount of volatile substances, which are released into the air from food during chewing and reach the olfactory sensors in the nose. As direct injection of headspace suffers from incompatibility with capillary columns and only isolates the main components with relatively high volatility, these problems can best be surmounted by the use of porous polymer enrichment techniques.

In our study we have tried to take into account these major problems described earlier. The complete analysis consists of headspace enrichment on Tenax of volatiles released during disintegration of the fruits, high resolution gas chromatography, mass spectrometric identification, quantification on a flame ionization detector and specific analyses of sulphur containing compounds. The instrumental analyses were related to those sensory analyses that showed significant differences. In the case of strawberries significant different sensory results could be obtained by comparing different varieties that were available at the same time. When dealing with Golden Delicious apples, the decrease of flavor quality as a function of controlled atmosphere storage is well known by the consumer.

From the point of view of component identification and organoleptic evaluation cultivated and wild strawberries have been intensively studied by different authors. Recently a summary of the literature has been given by Pyysalo *et al.* [2]. From the different studies it is clear that esters, alcohols, and carbonyls are important for the fruity and green notes of the strawberry flavor. Furthermore, 2,5-dimethyl-4-hydroxy-3-(2H)-furanone and 2,5-dimethyl-4-methoxy-3(2H)-furanone are

Table 1. Compared strawberry varieties and significant differ

Compared varieties	Sampling dates		
	3rd May '77 (greenhouse grown)	31st June '77 (plastic funnel)	25th July '77 (field grown)
Confitura	H (5%)	X	X
Gorella	L (1%)	X	X
Induka	X	X	X
Jesco	X	H (1%)	X
Primella	X		X
Senga Gigana	X		
Sivetta	X	X	X
Red Gauntlet	X	X	X
Souvenir de Charles Machiroux			
Bel Rubi			
Geko			
Domanil			
Hapil			

H, significant higher flavor intensity; L, significant lower parentheses.

ences in flavor intensity obtained by the ranking procedure.

(cultivation technique)			
27th June '78 (field grown)	*21st June '79 (field grown)*	*3rd July '79 (field grown)*	*12th July '79 (field grown)*
L (5%)	*L (1%)*	*X*	*X*
	H (1%)		
X	*X*		*X*
		X	
H (1%)		*H (5%)*	*X*
X	*X*		
X			
		X	
		L (5%)	

flavor intensity; (%), the significance level is given between considered to be important character impact compounds [2,3]. Recently the importance of sulphur containing compounds for the strawberry flavor has been recognized by the authors [4]. In apples, over 250 volatile components have been identified [5]. According to Flath [6] the components directly associated with the characteristic Delicious applelike aroma are: ethyl 2-methylbutyrate, n.hexanal and trans-2-hexanal. By a combined use of sensory and analtyical procedures the importance of 4-methoxyallylbenzene for the spicy note of apple flavor was established by Williams and coworkers [7].

EXPERIMENTAL

1. Sensory Analyses of Strawberry Varieties

Over the seasons 1977, 1978, and 1979, at several times different varieties of fresh ripe strawberries were organoleptically evaluated and analysed by GC-MS. The strawberries were field grown, greenhouse grown, or grown under plastic funnels. They were harvested at the ripe stage of maturation and were kept in the refrigerator at 5°C before taste testing and analyses. A survey of the different varieties together with indication of significant flavor intensity differences are given in Table 1.

For sensory testing scaling tests were performed [8]. The panel varied from ten to twelve members, who after sniffing and tasting, rated the fruits according to a 1 - 9 scale for flavor intensity: 1: extremely weak strawberry flavor; 9: extremely strong strawberry flavor. The panelists were blindfolded in order to madk the rather important appearance differences. As it is very difficult to compare taste characteristics of products which have been tasted with important time intervals significant differences could only be determined between varieties availabel at the same moment. Therefore, the individual scores for flavor intensity were transformed into ranknumbers and statistical analysis was performed by summing the ranknumbers and comparing the ranksums to the totals required for statistical significance. It should also be stressed that in order to exclude individual preferences the panelists ranked the samples on the basis of typical strawberry flavor intensity and not for flavor quality. However as nowadays the problem of the commercially cultivated varieties is their "watery" and tasteless character, there would probably be a good relation between strawberry flavor intensity and flavor quality.

2. Storage of Golden Delicious Apples

Golden Delicious apples, which represent 70% of the Belgian production, can be stored in controlled atmosphere from October until June. It is trongly suspected that during this storage period flavor quality decreases importantly. However, this is difficult to prove organoleptically as sensory analyses of samples with important time intervals are very difficult to perform. In order to advise agricultural consultants to make regulations for shorter storage periods and better quality, we wanted to prove this on an objective base. Therefore, the evolution of aroma components of Golden Delicious apples ripened in standard conditions after removal from controlled atmosphere at several times during the storage period were performed. The storage conditions were: controlled atmosphere at a temperature of 0.05°C and relative humidity of 93%. Ripening after removal from controlled atmosphere was performed at 20°C and relative humidity of 67-70%. At regular times, depending on the degree of alteration, aroma analyses were performed during the following ripening periods:

period I:	from	4th December	'78	to	15th December	1978
period II:	from	18th December	'78	to	12th January	1979
period III:	from	19th January	'79	to	2d February	1979
period IV:	from	6th March	'79	to	5th April	1979
period V:	from	9th April	'79	to	1st May	1979
period VI:	from	9th May	'79	to	20th May	1979.

3. Sampling Procedure and Sample Injection of Volatiles

For correlation with the falvor intensity scores the occurrence and concentration of volatiles, representatively isolated from the food matrix, was determined. In order to isolate a fresh representative flavor mixture, the volatiles liberated while macerating the strawberries and apples, were adsorbed on Tenax GC 60/80. The sampling procedure has been described before [9].

Description of sampling apparatus. The sampling apparatus is presented in Fig. 1.

A commercial blender A (Braun MX 32, Germany), in which all plastic parts have been changed to Teflon, was adapted with a glass flange B and fitted with a three-necked flange cover. During sampling helium gas was supplied through C. The cover was fitted with a splash head D, with dump valve E, which was necessary to exchange the contaminated air inside the blender (volatile organic air pollutants) by helium before sampling. The splash head E was connected to the adsorber F.

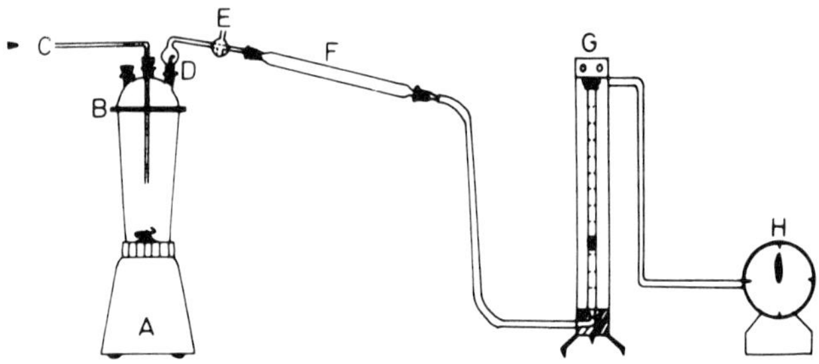

Fig. 1. Sampling apparatus.

The helium gas flow, regulated by a fine metering valve, was measured by the flow meter G and the total volume of the gas sample by the wet-testmeter H.

For its thermal stability and ability not to absorb water, a polyaromatic polar polymer, Tenax GC 60/80 mesh (Applied Science Lab., Inc., State College, Pennsylvania) was selected as an ambient temperature adsorbent and was packed in a relatively large glass adsorption column (i.d., 1.6 cm; length, 10 cm) with an adsorbent content of 5 g. For quantitative analysis helium passed through the adsorption column with a flow rate of 500 ml/min during 10 min. Loaded columns were closed and could be stored for a long period without loss of sample or changes in the composition.

About 250 g of fruits were placed into the blender (stalks were removed and apples were quartered), which was closed as hermetically as possible and provided with the features described above. Before disintegration was started, the inside space of the blender was washed with helium through dump valve E for 10 min. After switching valve E, disintegration of the fruits was started slowly (the rotation speed of the propellers was regulated by a Vareac), and adsorption was continued for 10 min, while further disintegrating the material.

Sample injection. The flavor components were desorbed in a helium stream (flow 20 ml/min) and transferred to a cooling spiral. The sample injection system is shown schematically in Fig. 2. For sample desorption the adsorption tube was placed in a separate oven (gradually warmed up to 220°C), which was connected to a two-position 6-port, high temperature injection valve (Valco Instruments Co., Houston, Texas), also by 1/16" thermically isolated zero volume fittings. The injection valve was constantly kept at 250°C in a separate oven.

POSITION 1 :

Transfer of volatiles to cooling spiral

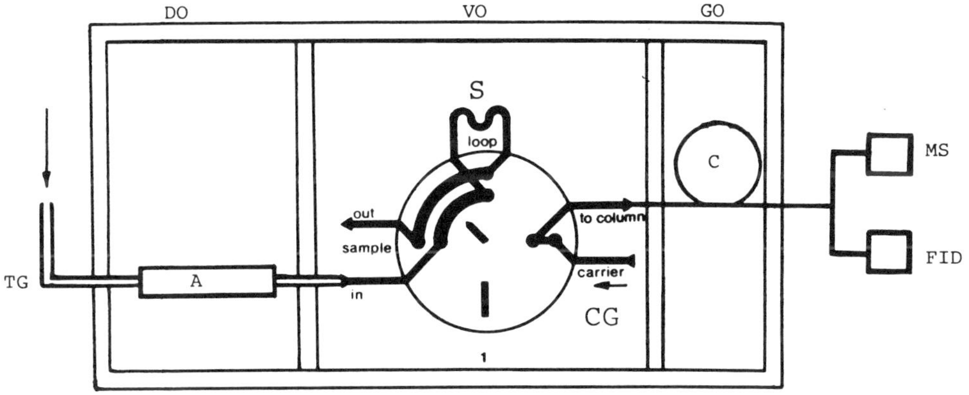

POSITION 2 :

Injection

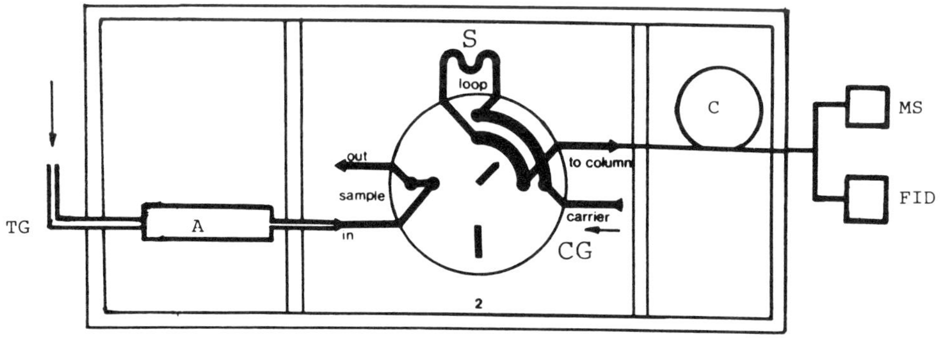

DO :	Desorption oven	A :	Adsorber
VO :	Valve oven	S :	Cooling spiral
GO :	Gaschromatographic oven	CG :	Carier gas
TG :	Transfer gas	C :	Column

Fig. 2. Injection system.

In transfer position the volatiles were desorbed from the Tenax adsorber and collected in a trap cooled with liquid nitrogen. For injection the liquid nitrogen was replaced by an injection oven while the 6-port injection valve was switched from collecting position to injection position.

4. Gas Chromatography

The volatiles from the different samples were separated on a Varian 3700 gas chromatograph, equipped with a flame ionization detector and a 200 m × 0.6 mm i.d. glass column coated with SE-30. Further operating conditions were: linear temperature programming from 20°C to 220°C at 2°C/min; carrier gas N_2, 3 ml/min; temperatures: injection valve 250°C, detector 250°C.

Quantitative determination of the different volatiles was performed by electronic integration (Infotronics 204) and by relating the peak intensities to the mean of the intensities of n.dodecane and n.tridecane as internal references. Therefore, before desorption 1 µl of a 1% solution of the n.alkanes in carbondisulphide (1 µg of each hydrocarbon/µl carbondisulphide) was injected on the adsorption column for quantization.

Besides analyses on a universal FID-detector for some strawberry varieties also specific analyses for sulphur compounds on a Tracor 560 gas chromatograph equipped with a flame photometric detector were performed. The Tracor 560 gas chromatograph was equipped with a 125 m × 0.6 mm i.d. capillary coated with SE-30. The flame photometric detector is 100-1000 times more sensitive for sulphur compounds than the FID-detector.

5. Gas Chromatogrpahy—Mass Spectrometry

The GC-MS apparatus consisted of a Varian 2700 gas chromatograph linked to a MAT 112 mass spectrometer. The mass spectrometer was directly coupled to the gas chromatograph by means of a glass tubing capillary (length 50 cm; 0.133 i.d.) heated at 250°C and allowing a constant flow of 2 ml of effluent into the MS source. In order avoid distrubance of the monitogram due to the water content of the sample, an effluent splitter was installed to send about 50% of the total effluent to a FID-detector. Operating conditions for GC-MS were: 250 m × 0.6 mm i.d. glass column, coated with OV-1; linear temperature programming from 20°C to 220°C at 2°C/min; carrier gas He; 2.8 ml/min and make-up to 4 ml/min before FID-MS splitter and to 30 ml/min before FID-detector; temperatures: injector 250°C; interconnecting lines 250°C; io source 250°C; ion source pressure 10^{-5} mmHg; trap current 300 µA, filament voltage 70 V, scan speed 100 masses/sec.

RESULTS

1. Strawberry—Comparison of Varieties

Quantitative GC-MS data. A typical gas chromatogram for the strawberry variety Sivetta obtained by adsorption on Tenax in standard conditions of the volatiles liberated while macerating the fruits is given in Fig. 3. The peak numbering is based on retention indices and corresponds with the indices in Table 2. By relating the electronic integration of the different aroma compounds to the intensity of 1 μg hydrocarbon (mean of n.dodecane and n.tridecane) relative intensities were calculated and listed in the tables. As an illustration the relative intensities expressed as μg/5 l headspace of the liberated flavor components of the sampling of 27th June 1978 are given in Table 2.

As shown in Table 2 strawberry flavor is composed of a whole range of volatile organic esters and some aldehydes (trans-2-hexanal and n.hexanal). A few sulphur containing compounds (dimethyldisulphide, methylthiol butyrate and methylthiol acetate), are present in lower concentrations in the FID-chromatogram and can only be distinguished in some varieties.

A complete representation of the quantitative GC data for the different sampling dates is beyond the scope of this presentation. However from overlooking these different tables it can be concluded that in some cases quantitative data can be used for a variety characterization, e.g., Jesco is the only variety with a higher amount of butyl butyrate as compared to the neighboring ethyl hexanoate. Also the 2-alkanones, 2-pentanone, and 2-heptanone are typical compounds and occur only in the varieties Souvenir DCM and Domanil. From the quantitative GC data it is also obvious that organic esters are the qualitatively and quantitatively most important class of volatiles in strawberry flavor. The main components are formed from volatile organic acids with an even carbon number (acetic, butyric, and hexanoic acid) and lower concentration esters are formed from propionic, isobutyric, isopentanoic, 2-methylbuteric, isohexanoic or 2-methylpentanoic acid. The aldehydes, n.hexanal, and trans-2-hexanal, which are formed enzymatically from linol- and linolenic acid [10] during disintegration, show important concentration fluctuations. Separate studies, in which the evolution of the volatiles as a function of maturation on the plant and after harvesting at immature stages was followed, indicated indicated that both aldehydes cannot be considered as variety characteristic but are connected with the maturity stage and are important for the "green" character in strawberries.

Table 2. Relative intensities of strawberry flavor components of the sampling of 27th June 1978

Flavor components	*$Ip\text{-}SE_{30}$*	*Souvenir*	*Bel Rubi*	*Geko*	*Sivetta*	*Gorella*
Ethyl acetate (a)	605	0.573	6.989	2.996	6.583	2.017
Methyl proprioate (a)	621	0.186	0.459	0.651	1.620	1.693
Isopropyl acetate	648	0.400	0.272	0.888	0.545	0.894
Methyl isobutyrate	670	-	-	0.070	0.133	0.125
2-Pentanone	670	0.349	-	-	-	-
Methylthiol acetate	675	0.156	0.134	-	-	-
Ethyl propionate	695	-	-	-	-	-
Propyl acetate	697	-	-	-	-	-
Methyl butyrate	706	11.086	27.428	40.055	24.782	7.794
Dimethyldisulphide	731	0.233	0.134	0.120	-	0.030
Ethyl isobutyrate	743	-	-	0.099	-	-
Isobutyl acetate	758	-	-	0.073	0.181	0.118
Methyl isopentanoate	761	0.109	2.192	-	1.156	-
Methyl 2-methylbutyrate	763	-	-	1.173	1.156	0.812
n.Hexanal	780	0.137	0.132	0.058	2.745	0.094
Ethyl butyrate	785	0.850	6.315	2.326	2.023	0.164
Butylacetate	796	-	3.526	2.740	1.723	0.423
Methyl pentanoate	806	0.124	0.114	0.189	0.016	0.021
Isopropyl butyrate	825	0.759	0.665	2.090	0.458	0.121
Trans-2-hexenal	831	0.238	0.185	0.083	6.434	0.077

Ethyl 2-methylbutyrate	836	-	0.217	0.127	-	0.054
Ethyl isopentanoate	838	-	0.720	0.222	-	0.037
Isoamyl acetate	861	0.422	0.595	0.424	0.312	0.390
2-Methylbutyl acetate	863	0.228	0.616	0.411	0.533	0.280
Methylthiol butyrate	871	0.333	-	-	-	-
2-Heptanone	873	0.459	-	-	-	-
Methyl isohexanoate	875	-	0.086	0.148	0.303	0.162
Propyl butyrate	884	-	0.273	0.133	-	-
Butyl propionate	892	-	0.115	-	-	-
Amyl acetate	896	-	0.080	-	-	-
Isopent-2-enylacetate	904	-	-	-	-	-
Methyl hexanoate	906	0.580	3.498	6.292	1.879	1.891
Isobutyl isobutyrate	938	-	0.155	-	-	-
Isohexyl acetate	967	-	0.078	-	-	-
Butyl butyrate	977	0.100	0.264	0.305	0.078	-
Ethyl hexanoate	981	0.174	0.710	0.305	0.065	-
cis-3-Hexenyl acetate	987	-	0.161	-	-	-
Hexyl acetate (a)	996	0.226	1.893	0.641	0.316	-
Isopropyl hexanoate	1021	-	0.053	-	-	-
Limoneen	1021	0.158	-	-	-	-
Isobutyl valerate		-	0.061	-	-	-
Sum		17.870	58.380	62.619	51.938	17.197

(a), Contaminated peak

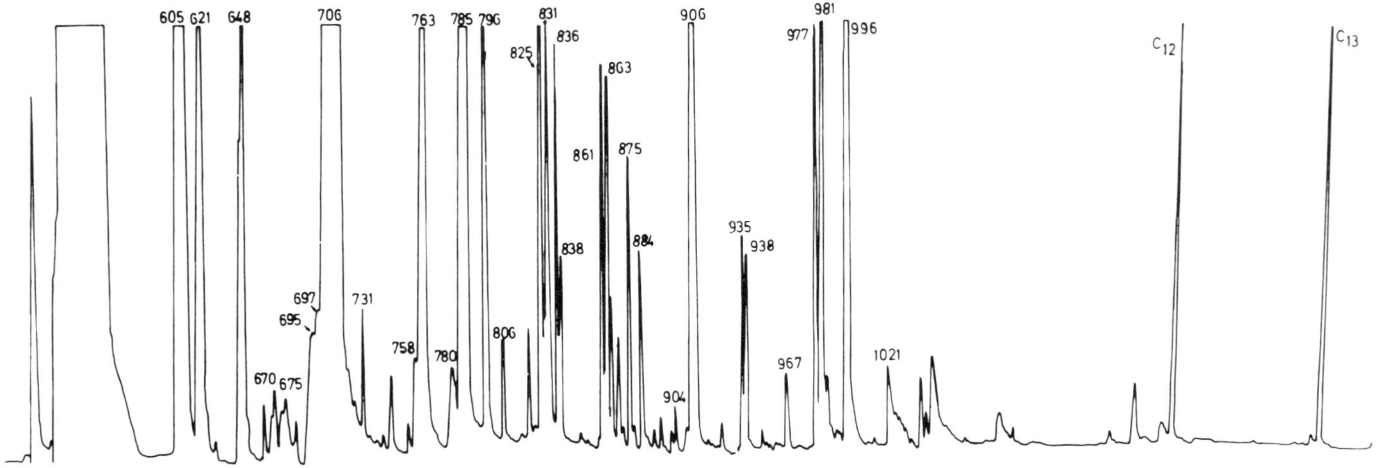

Fig. 3. Typical gas chromatogram of the variety Sivetta. Peak numbering by means of retention indices.

Relation of flavor quality and volatile composition. As a complete presentation of the quantitative GC data is not possible due to intensiveness, in Table 3 only the sum of the concentrations of volatiles released when macerating the fruits is given for the different varieties and sampling dates. Between brackets the results of the sensory analyses (significant more or less flavor intensive varieties per comparison) are repeated. As can be seen from Table 3, in the season 1977 there is a good relationship between flavor intensity scores and the sum of liberated strawberry volatiles. This was no longer the case for the sampling of 27th June 1978, where Souvenir, an "older" variety with excellent organoleptic reputation, was included in the comparison. Overlooking the individual aroma compositions and especially, comparing Souvenir DCM and Gorella, which had about an equal amount of released volatiles but were respectively significant more and significant less flavor intensive than the other varieties, showed some striking differences: occurrence of 2-pentanone, 2 heptanone, and of higher concentrations of sulphur containing compounds (methylthiol acetate, dimethyl disulphie, and methylthiol butyrate) in the FID-chromatogram of Souvenir. The importance of the sulphur compounds for the strawberry flavor was established by the following considerations. First, sulphur compounds have generally very low thresholds values and are important contributors in different foodstuffs [11, 12, 13]. Second, an organoleptic perception supported the importance of the sulphur compounds for the flavor intensity of Souvenir. When disintegrating the Souvenir strawberries in a Waring Blender and sniffing the headspace an intense "agreeable rotten odor" was observed, which only after a few minutes was dominated again by the fruity character of the esters. Third, as the sulphur compounds occured only in relatively small concentrations compared to the esters, specific analyses for sulphur containing compounds on a flame photometric detector were performed. As an illustration, the FPD-chromatogram taken in analogue isolation and GC conditions for Souvenir and Gorella are given in Fig. 4. As can be seen from the chromatogram besides methylthiol acetate, dimethyldisulphide, and methylthiol butyrate also methanethiol and dimethylsulphide can be observed in the FPD-chromatogram.

2. *Apples—Evaluation of Flavor Quality as a Function of Storage in Controlled Atmosphere*

A typical gas chromatogram of Golden Delicious apples is shown in Fig. 5. The peak numbering is related to the identifications in Table 4, which show an important amount of volatile esters, some alcohols, some aldehydes, and estragol

Table 3. Sum of liberated strawberry volatiles for the different varieties and sampling dates, which showed significant sensory results.[a]

Variety	Sampling dates				
	3rd May 77	31st June 77	27th June 78	21st June 79	3rd July 79
Confitura	330 (H)	273	-	-	-
Gorella	49 (L)	21	17	59 (L)	41
Induka	17	-	-	-	-
Jesco	148	930 (H)	-	335 (H)	-
Primella	120	-	-	-	-
Senga Gigana	48	-	-	-	-
Sivetta	170	77	52	238	-
Red Gauntlet	-	43	-	-	39
Souvenir de Charles Machiroux	-	-	18 (H)	-	161 (H)
Bel Rubi	-	-	58	241	-
Geko	-	-	63	-	-
Domanil	-	-	-	-	360
Hapil	-	-	-	-	27 (L)

[a]Expressed as μg/5 l headspace.
(H), Sum of volatiles with significant higher flavor intensity scores
(L), Sum of volatiles with significant lower flavor intensity scores

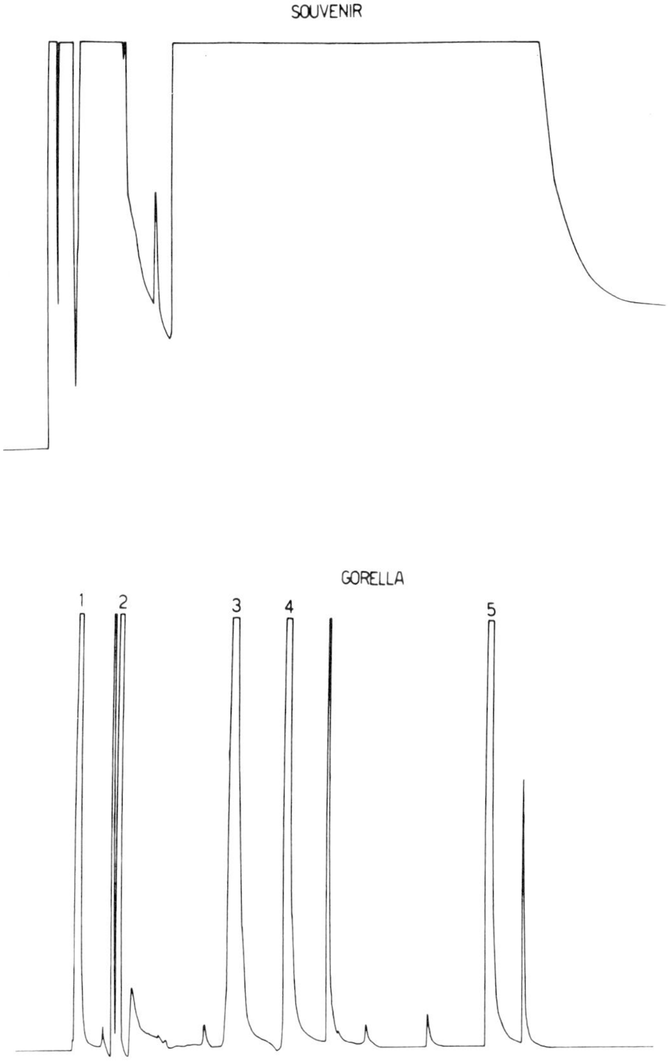

Fig. 4. Specific analyses for sulphur containing compounds 1, Methanethiol; 2, Dimethylsulfide; 3, Methylthiol acetate; 4, Dimethyldisulfide; 5, Methylthiol butyrate.

(4-methoxyallylbenzene). Undecane and tridecane are internal standards. The C-6 aldehydes hexanal and trans-2-hexenal are formed enzymatically by oxidative cleavage of linolic and linolenic acid during disintegration of the fruits [14]. As to the biogenesis of the esters it is accepted that they are

Table 4. Identified compounds in Golden Delicious apples.

1. Acetaldehyde	*13. Butyl propionate*
2. Propanol	*14. Pentyl acetate*
3. Ethyl acetate	*15. Butyl butyrate*
4. Butanol	*16. Hexyl acetate*
5. Propyl acetate	*17. Butyl 2-methylbutyrate*
6. 2- or 3-methylbutanol	*18. Pentyl butyrate*
7. 2-methylpropyl acetate	*19. Undecane*
8. Hexanal	*20. Hexyl propionate*
9. Butyl acetate	*21. Hexyl butyrate Butyl hexanoate*
10. Trans-2-hexenal	*22. Estragol*
11. 3-methylbutyl acetate	*23. Hexyl 2-methylbutyrate*
12. Propyl butyrate	*24. Tridecane*

synthetized enzymatically by combining acid and alcohol moieties [15]. The straight chain acids could be formed by β-oxidation of the straight chain fatty acids and the branched chain acid moieties, such as 2-methylpropionates, 2-methylbutyrates or 3-methylbutyrates could be derived from amino acids (valine, isoleucine, and leucine).

As an illustration of the quantitative GC data in Table 5 the evolution of Golden Delicious volatiles (isolated by the headspace concentration technique) as a function ripening in standard conditions after removal from controlled atmosphere is given for period II (ripening from 18th December 1978 to 12th January 1979). Overviewing the complete data of the different ripening periods indicates that if the sum of the volatiles shows a maximum, a maximum is also found for most of the individual compounds also. However examination of these data shows that for the longer storage ripening experiments hexanal becomes more important in relation to the volatile organic esters, indicating that after long storage the green character of the apple is becoming more important.

From the point of view of physical measurement relationships the total sum of released volatiles during disintegration can be used as an aroma quality parameter. In Fig. 6 the evolution of the sum of apple volatiles, isolated in the headspace concentration technique, is given for the different

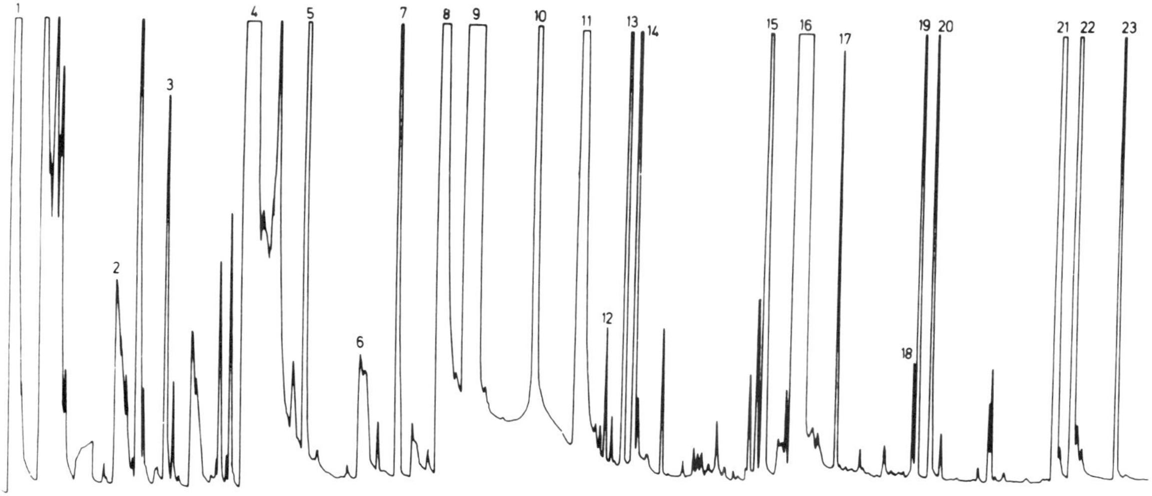

Fig. 5. Typical gas chromatogram of Golden Delicious apples.

Table 5. Evoluation of Golden Delicious volatiles (isolated by headspace concentration) as a function of ripening in standard conditions after removal from controlled atmosphere (period II: 18th December 78 to 12th January 79).

Compounds	—						
Propyl acetate	0.346	0.610	2.476	2.568	3.389	3.652	3.857
2-Methylpropyl acetate	1.963	2.377	2.926	2.286	1.551	1.221	1.178
Hexanal	5.503	3.740	15.767	13.261	11.880	11.053	9.743
Butyl acetate	21.555	18.383	46.102	59.283	73.988	54.136	47.489
Trans-2-hexenal	1.785	1.595	5.433	3.981	5.221	3.277	2.452
3-Methylbutyl acetate	4.056	4.537	6.173	9.923	11.000	10.328	11.928
Butyl propionate	0.440	0.424	1.324	1.781	1.576	0.886	1.045
Pentyl acetate	1.184	1.086	1.767	2.296	1.987	1.308	1.484
Butyl butyrate	1.480	1.060	2.276	3.290	3.036	2.629	2.727
Hexyl acetate	18.825	13.882	34.068	37.966	45.970	25.687	24.132
Cis-3-hexenyl acetate	1.148	1.519	1.730	1.801	0.997	1.763	0.607
Butyl 2-methylbutyrate	0.194	0.210	0.445	0.554	0.562	0.628	0.882
Pentyl butyrate	0.057	0.081	0.166	0.106	0.095	0.146	0.106
Hexyl propionate	0.240	0.240	0.644	0.707	0.588	0.380	0.552
4-Methylpentyl butyrate	0.074	0.138	0.307	0.242	0.113	0.202	0.192
Hexyl butyrate	2.160	1.520	3.451	4.760	4.278	3.403	3.641
Estragol	0.209	0.218	1.928	2.790	1.861	1.122	0.983
Hexyl 2-Methylbutyrate	0.216	0.248	0.693	0.810	0.785	1.150	1.059
Sum	61.4	51.8	127.7	148.5	164.9	123.0	118.0

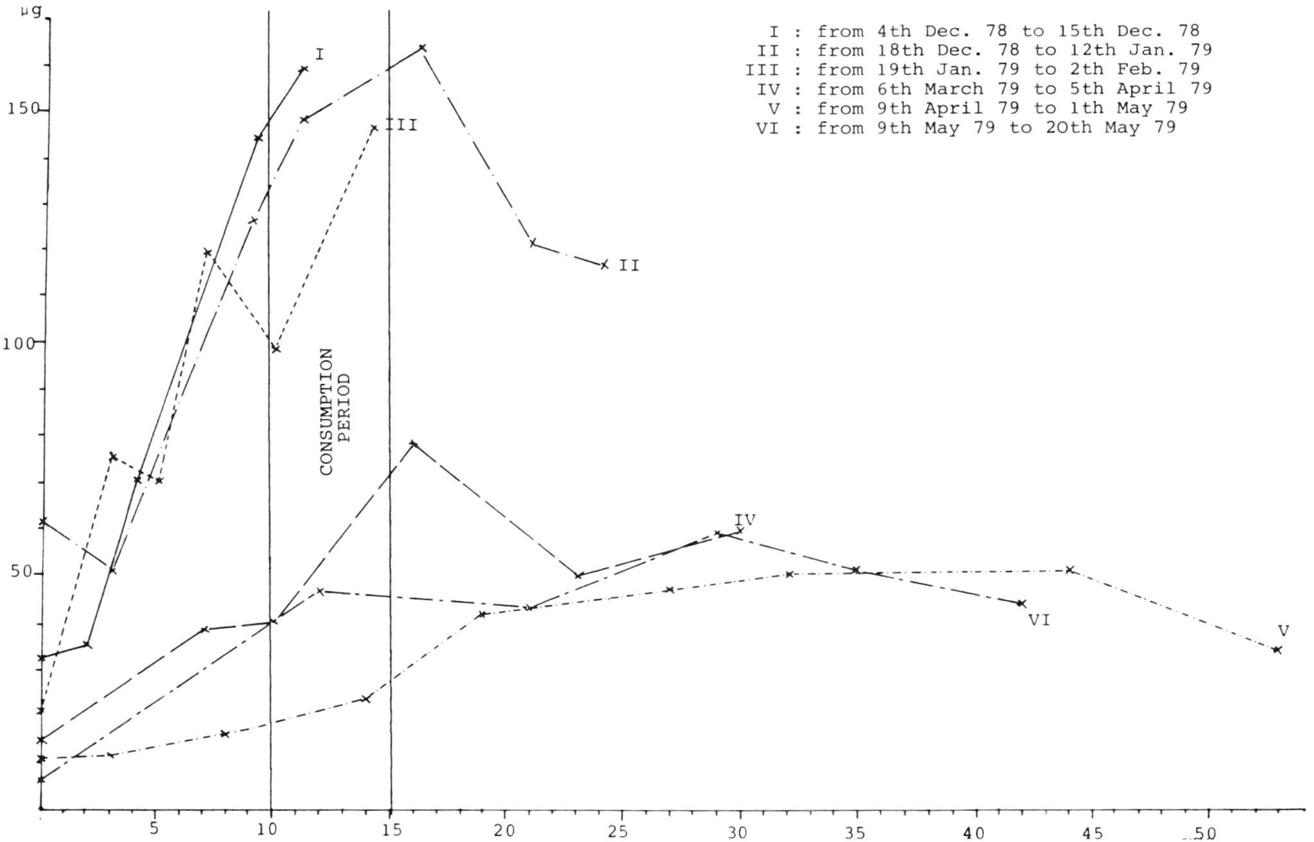

Fig. 6. Evolution of the sum of apple volatiles, isolated in the headspace condensation technique, for the different ripening periods.

ripening periods. In relation to eating quality, it is accepted that the consumption period is 10-15 days after removal from controlled atmosphere (distribution, shop storage, home storage). Examination of Fig. 5 shows a clear distinction between aroma evolution during ripening after short storage (until December and January) compared to long controlled atmosphere storage (March, April, May). Also these results prove objectively by physical measurements that although long stored apples have excellent appearance and are free of defects, their flavor quality has decreased importantly. In order to improve flavor quality and to avoid further complaints of consumers regulations should be established for shorter storage times.

CONCLUSION

As a conclusion we may state that objective measurement of fruit flavor quality is possible by measuring the volatiles released while macerating the fruits and by specific analyses of flavor important compounds as is the case for the sulphur containing compounds in strawberries. For representative isolation of volatiles, headspace concentration on Tenax of volatiles, released during fruit disintegration is a useful technique, which compared to other procedures also has the advantage of allowing a very fast isolation of volatiles. Although time-consuming, the most efficient procedure for determination of flavor important compounds is relating the GC compositions to significant sensory results and performing a lot of comparisons. From a practical point of view this study established that important differences in flavor quality occur between strawberry varieties and that especially the currently cultivated varieties have poor flavor quality. As to apple storage in controlled atmosphere, this study proves objectively that there is an important flavor decrease for apples, stored in controlled atmospheres for long periods. In order to offer an acceptable quality regulations for shorter storage periods and better quality should be established.

REFERENCES

1. Teranishi, R., Hornstein, I., Issenberg, P., and Wick, E. L., "Flavor Research, Principles and Techniques," Marcel Dekker Inc., New York (1971).
2. Pyysalo, T., Honkanen, E., and Hirvi, T., *J. Agric. Food Chem. 27,* 19 (1979).
3. Sundt, E., "Proceedings of the Scandinavian Symposium on Aroma Research," *Naeringsmiddelindustrien 23,* 5 (1970).

4. Dirinck, P., De Pooter, H., Willaert, G., and Schamp, N., *J. Agric. Food Chem. 29,* 316 (1981).
5. Van Strating, S., and de Vrijer, F., List of Volatile Compounds in Foods, Cent. Inst. Nutr. Food Res., Zeist, Netherlands (1973).
6. Flath, R. A., Black, D. R., Guadagni, D. G., McFadden, W. H., and Schultz, T. R., *J. Agric. Food Chem. 15,* 29 (1967).
7. Williams, A., Lea, A., and Timberlake, C., A. C. S. Symposium Series, No. 51, "Flavor Quality: Objective Measurement," R. A. Scanlan (Editor).
8. Kramer, A., *Chemical senses and flavour 1,* 121 (1974).
9. Dirinck, P., Schreyen, L., and Schamp, N., *J. Agric. Food Chem. 25,* 759 (1977).
10. Bonnet, J. L., and Crouzet, J., *J. Food Sci. 42,* 625 (1977).
11. Maga, J. A., *Crit. Rev. Food Technol. 5,* 153 (1975).
12. Maga, J. A., *Crit. Rev. Food Technol. 5,* 241 (1975).
13. Maga, J. A., *Crit. Rev. Food Technol. 6,* 147 (1976).
14. Drawert, F., Tressl, R., Heimann, W., Emberger, R., and Speck, M., *Chem. Mikrobiol. Technol. Lebensm. 2,* 10 (1973).
15. Salunkhe, D. K., and Do. J. Y., *Crit. Rev. Food Technol. 6,* 161 (1976).

AUTOMATIC DIRECT HEAD SPACE GC ANALYSIS OF FLAVORS WITH CAPILLARY COLUMN AND MULTIDETECTOR SYSTEMS

Pio Gagliardi

Giuseppe Romano Verga

Carlo Erba Strumentazione
Rodano, Milano, Italy

I. INTRODUCTION

The combined use of head space technique and high resolution gas chromatography is often essential for the determination of complex mixtures such as, for instance, the constituents of volatile aromatic compounds in natural products, food and beverages. In fact the head space vapours of these compounds are frequently a mixture of numerous heterogeneous volatile components of wide concentration, the trace presence of which may nevertheless be fundamental for the final evaluation of the product being analysed.

Analysis of such mixtures calls for a high separation efficiency gas chromatographic system; this can only be obtained using capillary columns.

ISBN 0-12-169080-6

Direct analysis of head space samples by high resolution gas chromatography can thus be very useful because it combines the characteristics of the head space method simplicity, high sensitivity and reduced analysis time with the excellent separation efficiency of capillary columns.

However, the use of these columns for direct analysis of head space gas samples may give rise to certain difficulties: these samples often consist of small amounts of volatile components diluted in large volumes of air. On account of the high analytical sensitivity required, large gas samples have to be introduced, thus limiting the separation efficiency of the column. The efficiency can only be preserved if the column is able to reconcentrate the broad inlet band from the injector into a narrower band.

These limitations can be overcome by enrichment techniques through the adsorption of volatile components on specific adsorbents (activated charcoal, porous polimers) and subsequent extraction with solvents or by thermal desorption (1,2,3,4,5). This can also be obtained by placing special accessories (traps) between the injector and the capillary colum (6-7).

Unfortunately these enrichment techniques call for long analysis time, do not altogether exclude the possibility of partial or even total loss of some components, and do not permit an easy automation of the analytical system which is particularly important in pollution tests and quality control.

In many cases rapid, direct analysis of head space samples can give sufficient information to establish a product's "fingerprint". In addition, quality control may only require qualitative and quantitative analysis of certain components, which can only be determined using selective detectors.

For these reasons, a true high resolution gas chromatographic system for head space analysis must maintain the separation efficiency of capillary columns while tolerating direct injection of relatively large gas samples; permit different selective detectors to be used even simultaneously, and the complete system to be run fully unattended.

The efficiency of the gas chromatographic system depends to a large extent on the use of capillary columns with specific characteristics: they must tolerate large sample volumes, show low catalytic activity and high inertness while ensuring a low bleed rate.

Large sample capacity can be achieved with a thicker stationary phase film (8) which not only permits larger sample volumes to be injected but also facilitates re-concentration of the inlet band in the inlet section of the column. Thick film capillary columns 1-2 microns are efficient and long-lasting and can be easily obtained using non-polar stationary phases, especially silicone gums. Problems are still persisting for the preparation of columns coated with high polarity stationary phases.

In such cases wide bore SCOT or WWCOT columns may be suggested. Lower efficiency of these columns compared to narrow-bore types can be partially compensated by their greater length, in view of their lower pressure drop (9-10-11).

The stationary phase must be suitable for subambient temperature operations. Sample injection at a low temperature followed by temperature programming is the best way of achieving good separation of the highly volatile fractions, peculiar of many head space samples, while ensuring easier recon - centration of the inlet band.

Low catalytic activity and high inertness minimize adsorption or catalytic activity of reactive components even at trace levels. Finally, low bleed facilitates temperature programming operations and the use of selective detectors, particularly electron capture detector (ECD).

The use of individual detectors, in a given order, prolongs operations unduly because they have to be changed, the analytical system has to be reconditioned each time, and analysis has to be repeated. A multidetector system is better in that it gives several items of information from a single gas chromatographic analysis, therefore drastically reducing analysis times. The simultaneous use of different detectors can also more easily provide analytical information which may be particularly useful when analysing complex mixtures in which specific trace components may be influenced by adjacent major sample components (12-13). In practice, therefore, multidetector operations can in these cases improve selectivity and increase the number of components detectable.

Finally an automated HSGC system, besides permitting the totally unattended analysis of a large number of samples, improves the reproducibility of quantitative data because the risk of manual errors in conditioning, sample withdrawal and injection is minimized.

II. INSTRUMENTATION

This paper proposes a high resolution gas chromatographic system for automatic head space analysis (Fig. 1) consisting of a capillary-dedicated 2900 Series gas chromatograph coupled with an HS Mod. 250 automatic sampler. The gas chromatograph

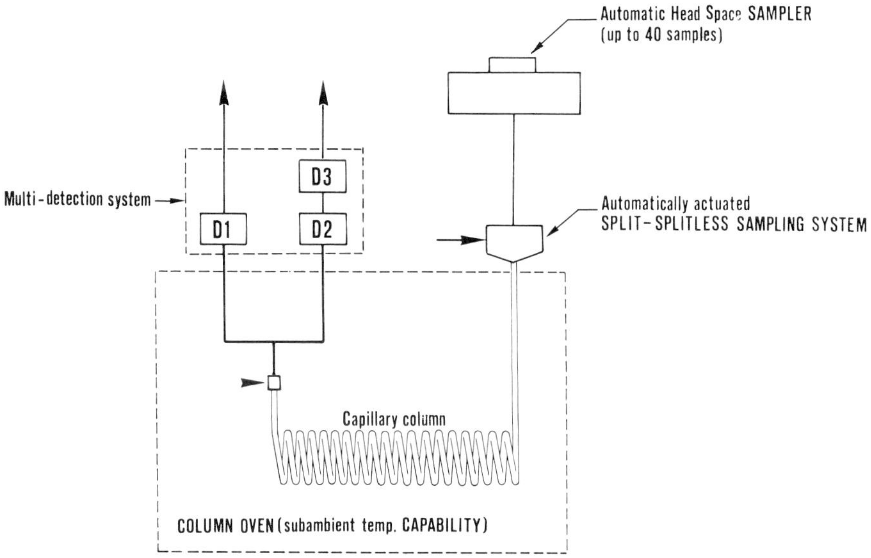

Fig. 1. Head space high resolution gas chromatograph configuration.

column oven can easily be operated at sub-ambient temperatures (down to 0°C). For this purpose a special heat exchanger is fitted, fed with liquid cooled by an external cryogenic unit.

The head space sample can be introduced automatically directly into the gas chromatograph injector in the split or splitless mode, as the relative valves are operated pneumatically from a suitable control module. This sampling system is completely independent of the capillary column inlet pressure and can therefore operate with all column types, including low pressure drop types such as the wide bore ones.

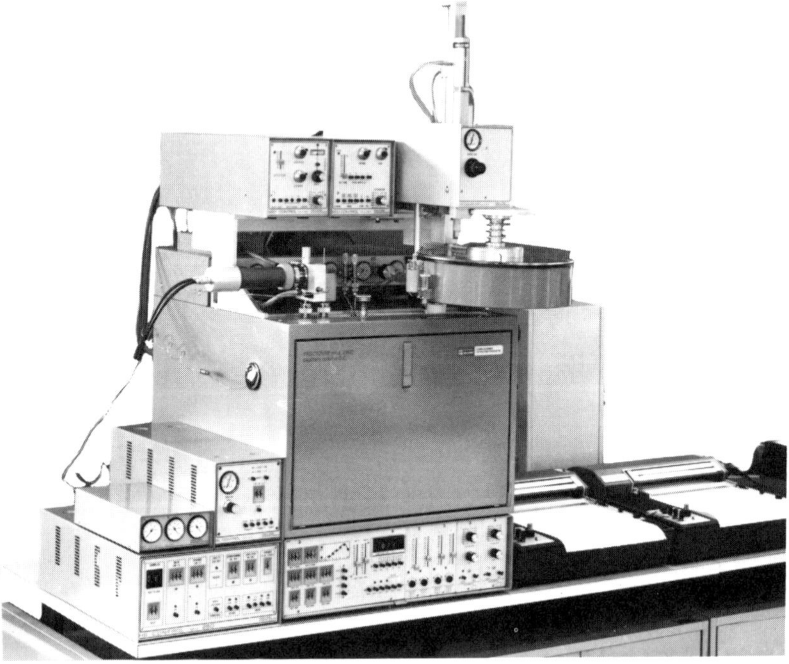

Fig. 2. GC instrumentation assembly.

Fig. 2 illustrates the assembly of this instrument. The automatic sampler, at the top of the gas chromatograph, consists of turntable housing up to forty 10 or 5 ml vials in a temperature controlled bath ranging from 30 to 120 $\pm$ 0.1°C and an electropneumatic system for sample introduction with a gas-tight syringe. The syringe is held at a constant temperature, independent of the vial temperature, between 30 to 120 $\pm$ 0.1°C and is flushed with ambient air before sampling to avoid memory effects.

The size of the gas sample withdrawn from each vial can be set between 0.1 and 2.5 ml. The controls regarding vial temperature, sampling syringe temperature, sample withdrawal and injection and analysis times are all grouped on a separate module.

The sampler can be fitted rapidly, without affecting access to the injection port for manual sampling, when necessary.

Multidetector Systems

The 2900 Series gas chromatograph is designed to permit multidetector configurations which can be set up easily by connecting standard detectors either in parallel or in series, or both.

Fig. 3 shows a multidetector system in which an FID is coupled in parallel with a nondestructive ECD which is in turn connected in series with an FPD.

The parallel FID-ECD connection is performed by a special 1:1 fixed ratio splitter. This is a glass lined tube deactivated before use with a solution of 0.1% carbowax 1000 in methylene chloride, then heated at 250°C in an inert gas stream. The splitter is connected to the capillary column with a special T-fitting providing an inlet for a make-up gas line.

The ECD-FPD connection in series is accomplished by an easily fitted adapter with a built-in line for auxiliary hydrogen for the flame photometric detector.

Make-up and auxiliary gas lines are already fitted on the GC, with independent channels for easy optimization of the various detectors.

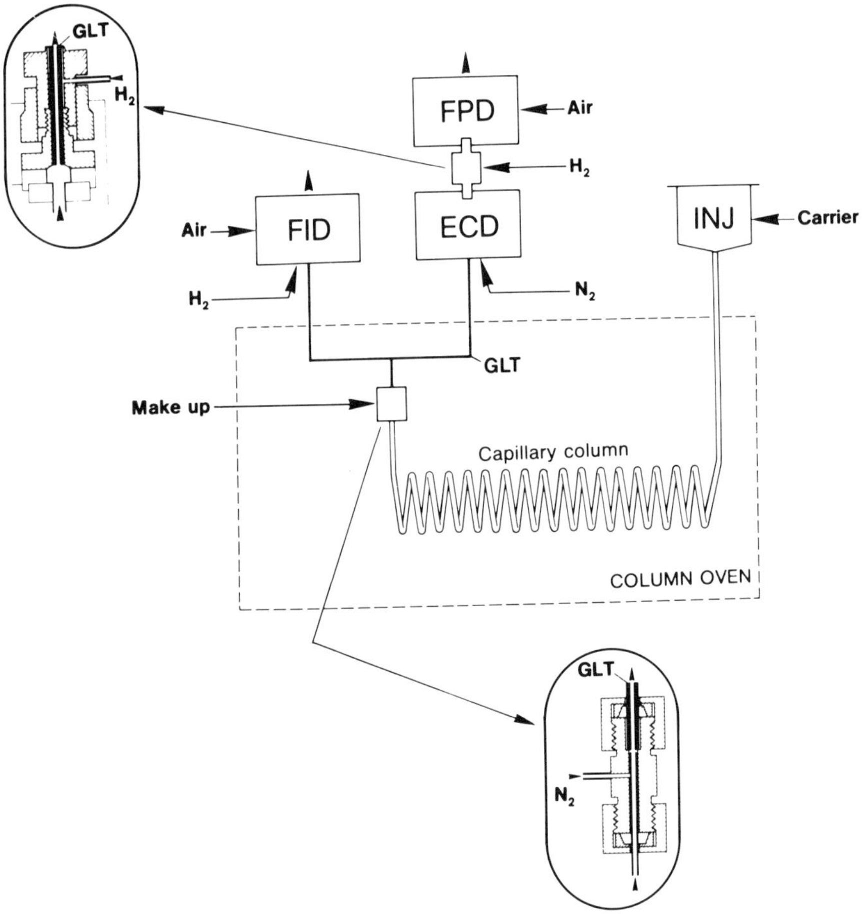

Fig. 3. Multidetector (FID-ECD-FPD) system configuration .

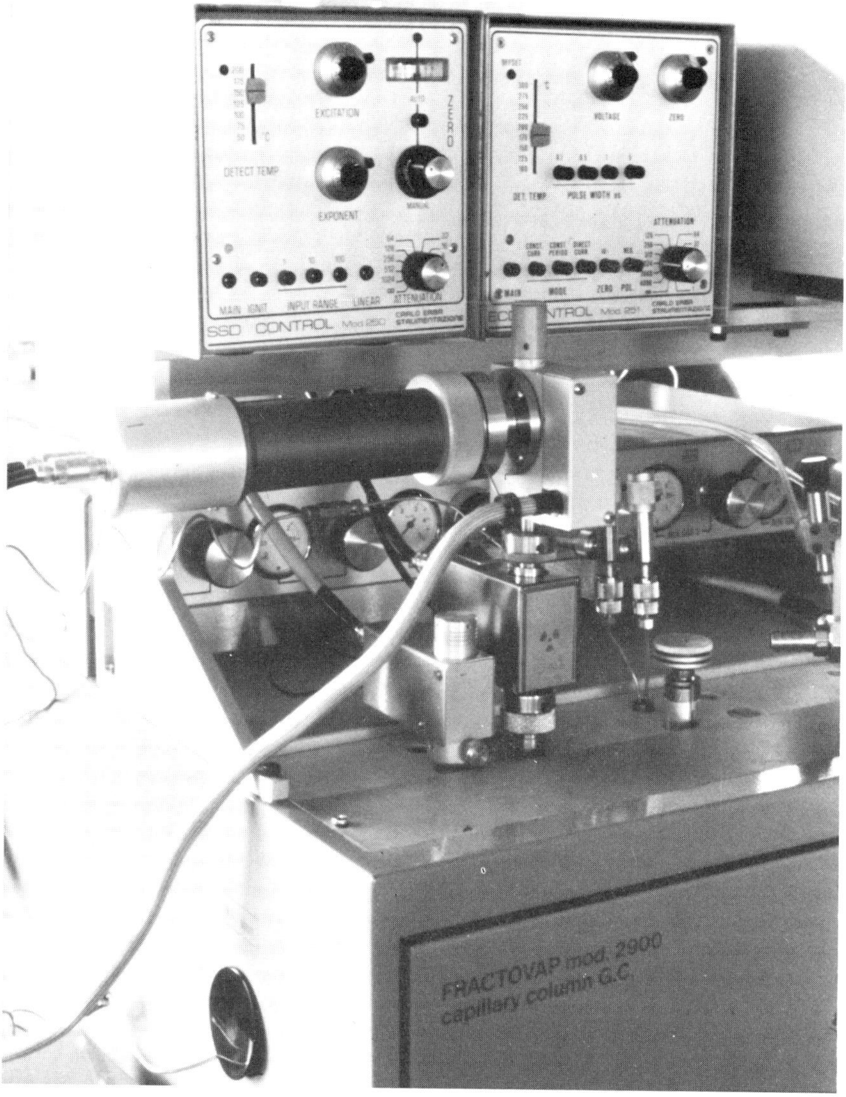

Fig. 4. Multidetector system (FID-ECD-FPD) assembly.

Figure 4 shows the practical assembly of this multidetector configuration on the 2900 Series gas chromatograph; access to the detectors is facilitated by the fact that the injection and detection compartments can be installed in different positions.

An additional feature (Fig. 5) is the large column oven, which allows easy fitting of the capillary column to the outlet splitter and the make-up gas line.

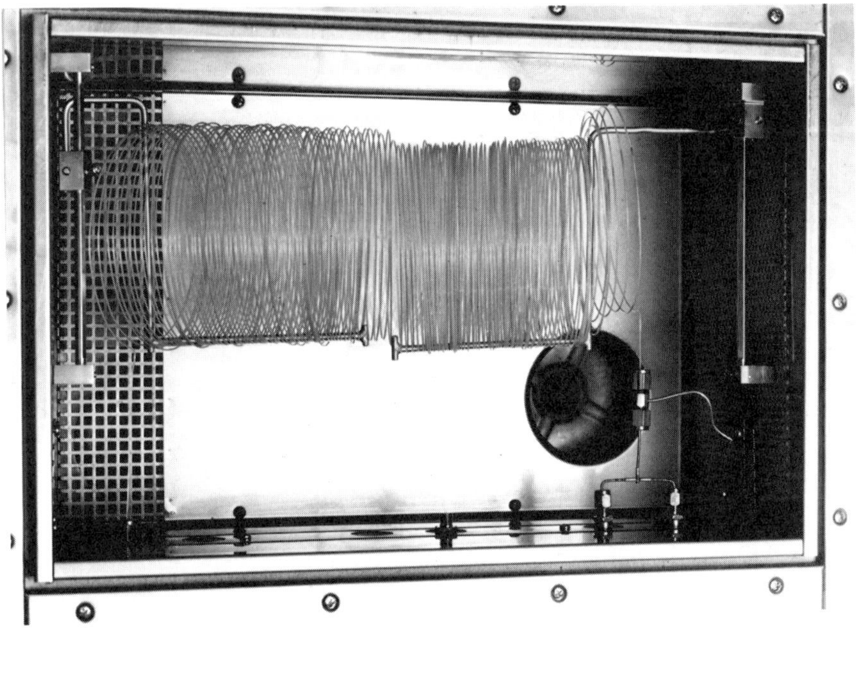

Fig. 5. GC oven with capillary column , GLT outlet splitter and make-up gas line.

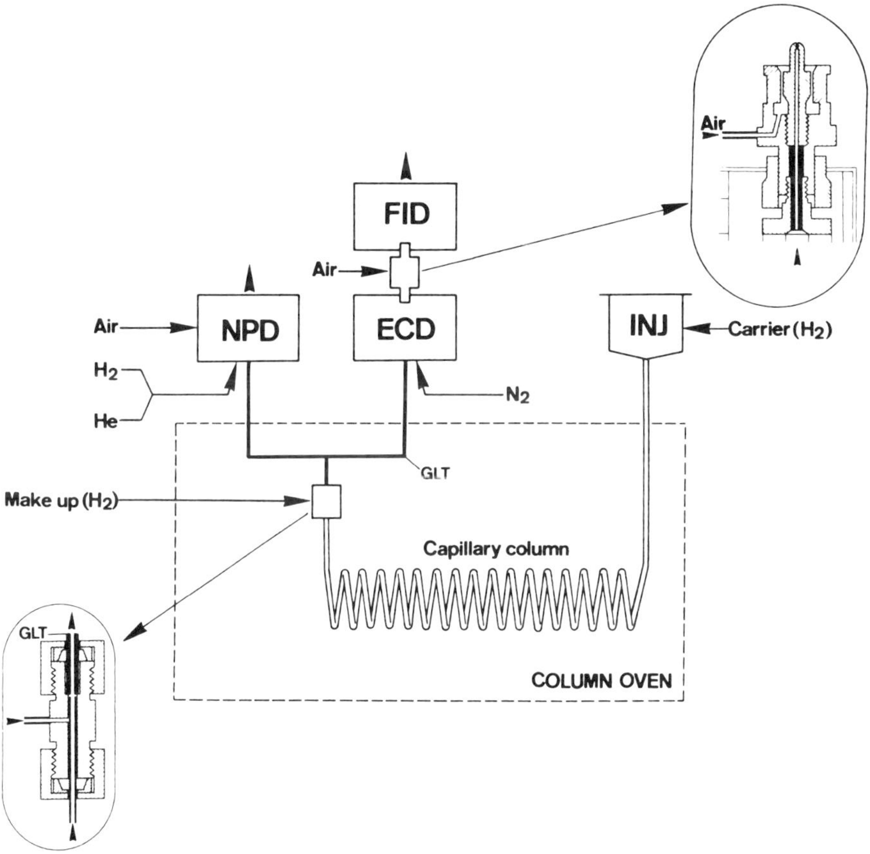

Fig. 6.' Multidetector (NPD-ECD-FID) system configuration.

A second multidetector configuration with NPD, ECD and FID is shown in Fig. 6. Here hydrogen for the FID flame is supplied through the make-up line at the capillary column outlet without limiting the ECD performancc. Helium also flows through the hydrogen line to the NPD, to optimize its response.

Simply at a flip of a switch the NPD Mod. 40 permits a choice of operating modes, N, P or NP, for the selective detection of nitrogen or phosphorus containing compounds. This will allow the response to purely carbon containing compounds to be suppressed or minimized while enhancing that to nitrogen or phosphorus containing compounds.

In this configuration the FID is connected in series to the ECD by means of an easily fitted adapter with built-in line for connection to the auxiliary air line for the FID.

III. APPLICATIONS

The analytical system proposed is suitable for multiple uses in various fields. Examples are analysis of volatile aromatic substances, environmental pollution and quality control in industry. This paper describes some specific applications of HS analysis with capillary columns and a multidetector system.

The analysis shown in Fig. 7 refers to samples of an aromatic herb from different origin. The dry herb was ground and conditioned at 60°C in a vial, after which the gas in the head space was analysed (14). In this case the FID showed significant qualitative differences between samples A and B, thus confirming the organoleptic assessment.

The second example refers to analysis of the head space gas in a beer sample; quality control requires determination not only of the main components such as aldehydes, esters and higher alcohols, but also of certain diketones and sulphur compounds (15-16-17-18).

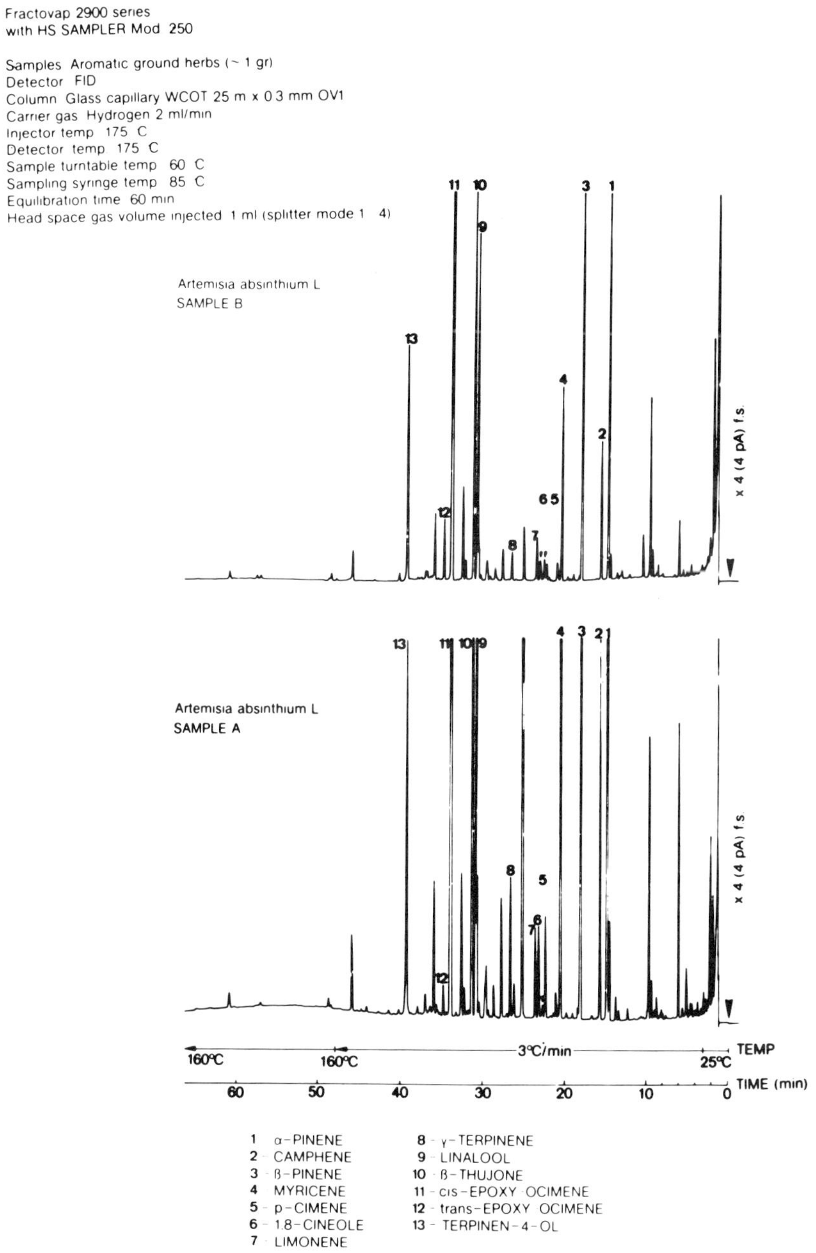

Fig. 7. Head space analysis of aromatic herbs.

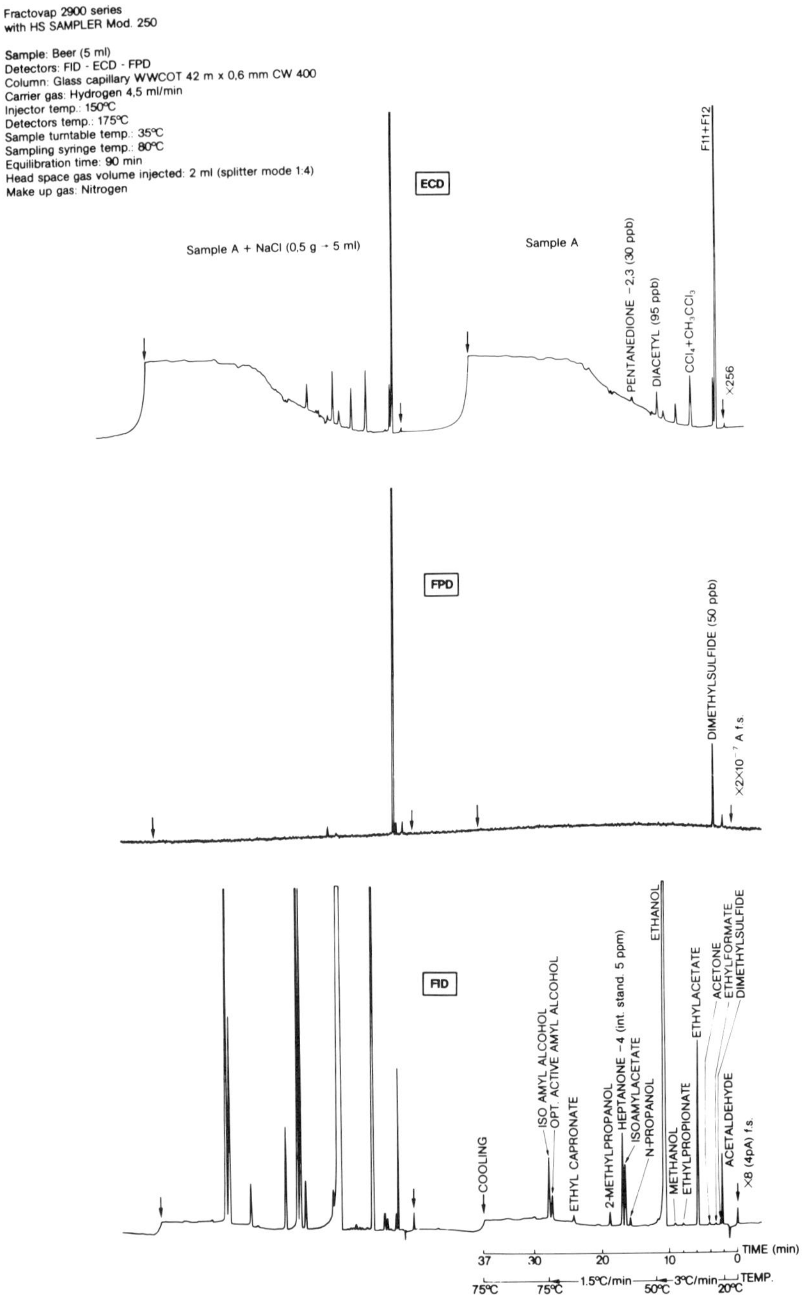

Fig. 8. Simultaneous responses of three different detectors to a beer sample head space.

Fig. 8 shows the simultaneous responses from the three detectors, FID, ECD and FPD, obtained analysing the head space in the beer sample both as such and NaCl-added to enrich the volatile components of the gaseous phase. Gas chromatographic analysis was performed by temperature programming from 20 to 75°C in about 30 min.

The type of column used (WWCOT with CW 400) (courtesy of Dr. P. Sandra - University of Gent, Belgium) not only featured complete separation of the main constituents, but also determined specific components such as diacetyl, 2-3 pentanedione, and dimethylsulphide present at concentrations of a few ppb. The relative standard deviation of quantitative data on 10 vials containing the same beer sample was ~ 5% and the response was linear between 0.2 and 2 mg l^{-1} for DMS and 2-3 pentanedione (19).

Fig. 9 refers to analysis of two coffee samples (raw and toasted beans) performed with the same type of capillary column and the same multidetector configuration. Ten whole beans were placed in each vial, which was then conditioned for three hours. The chromatograms show the large number of components detected in the toasted coffee using these three detectors. The possibility of working at a programmed temperature from 15°C permitted adequate separation even of the most volatile sulphur containing compounds.

Fig. 10 shows head space analysis of two samples of caramel colours. This was carried out at a programmed temperature of 25-85°C using a WCOT capillary column (MEGA) coated with 1.4 µm film thickness JXR methyl silicone. The FID and ECD responses show marked qualitative differences between the two samples, A and B.

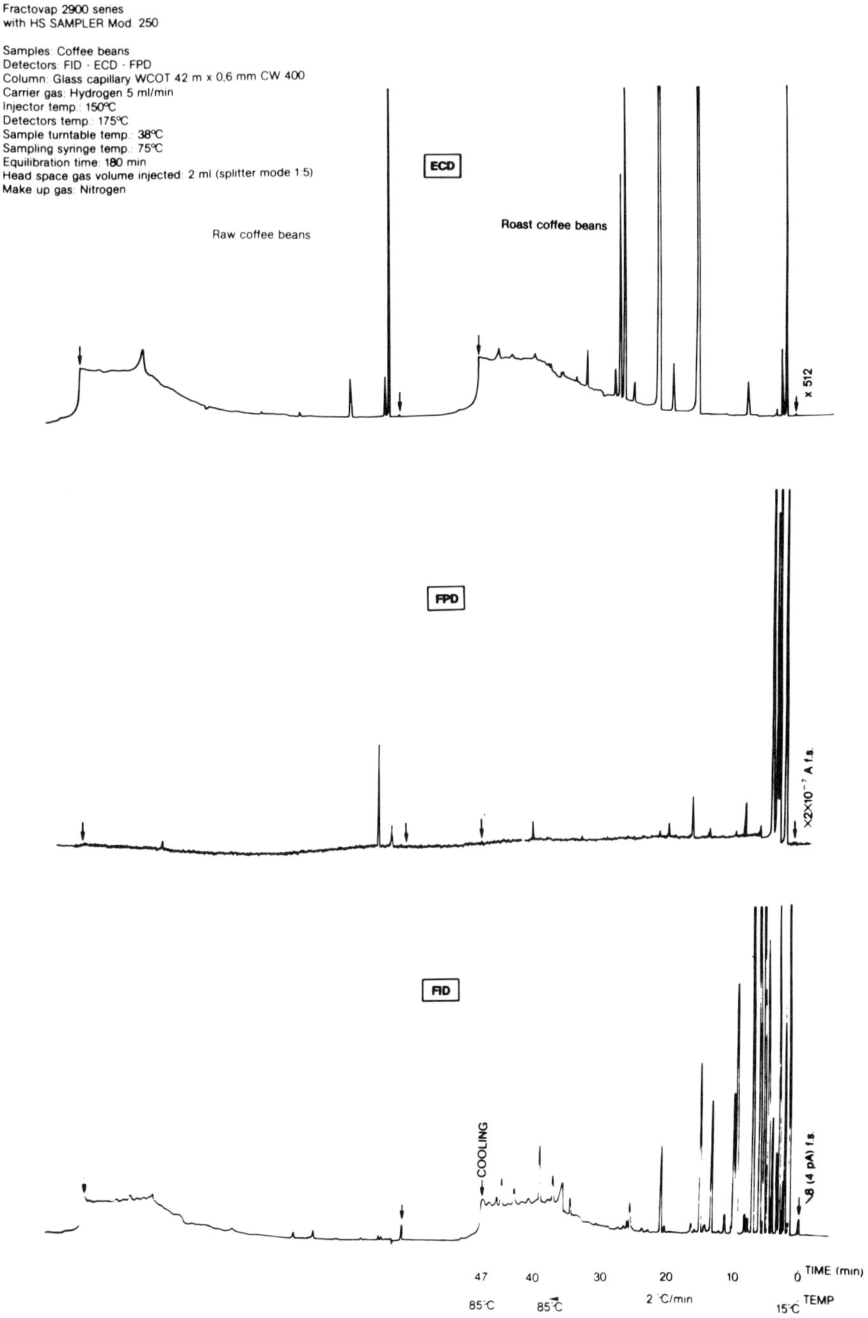

Fig. 9. Analysis of volatile constituents in roast and row coffee beans.

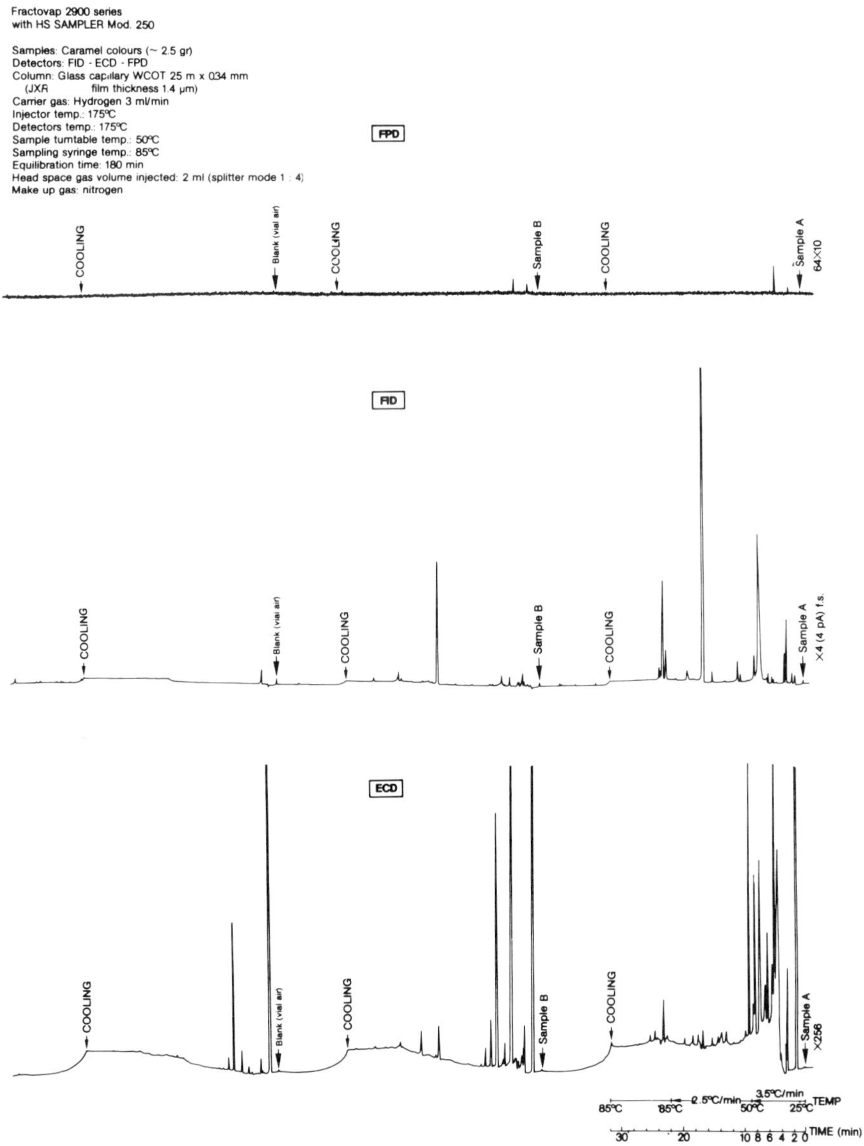

Fig. 10. HS analysis of caramel colours samples.

Analysis of samples of tobacco, shown in Figs. 11-12 were made with a WWCOT capillary column, but with different multidetector configurations. In the first case (FPD-FID-ECD) differences were found between samples A and B, particularly in the FID and ECD responses. In the second case (ECD-NPD-FID) the NPD shows clearly certain nitrogen containing compounds present in very limited amounts, even though there was a large number of other components, as shown by the FID. This analysis was done with the NPD in N mode which minimized the response to carbon.

Finally, Fig. 13 shows determination of acrylonitrile in olive oil, used as simulating fatty foodstuff migration liquid. This analysis imposes the use of the selective NPD in view of the high sensitivity required and the presence of the numerous compounds released by the olive oil after conditioning at 80°C (20). A thick film (JXR) capillary column was used for this analysis, at 40°C, and analysis time was about 2 min. In these analytical conditions minimum detectable amount of acrylonitrile in the olive oil was around 20 ppb, with good linearity of response, as shown in Fig. 14. Standard deviation on the quantitative data was better than 5% .

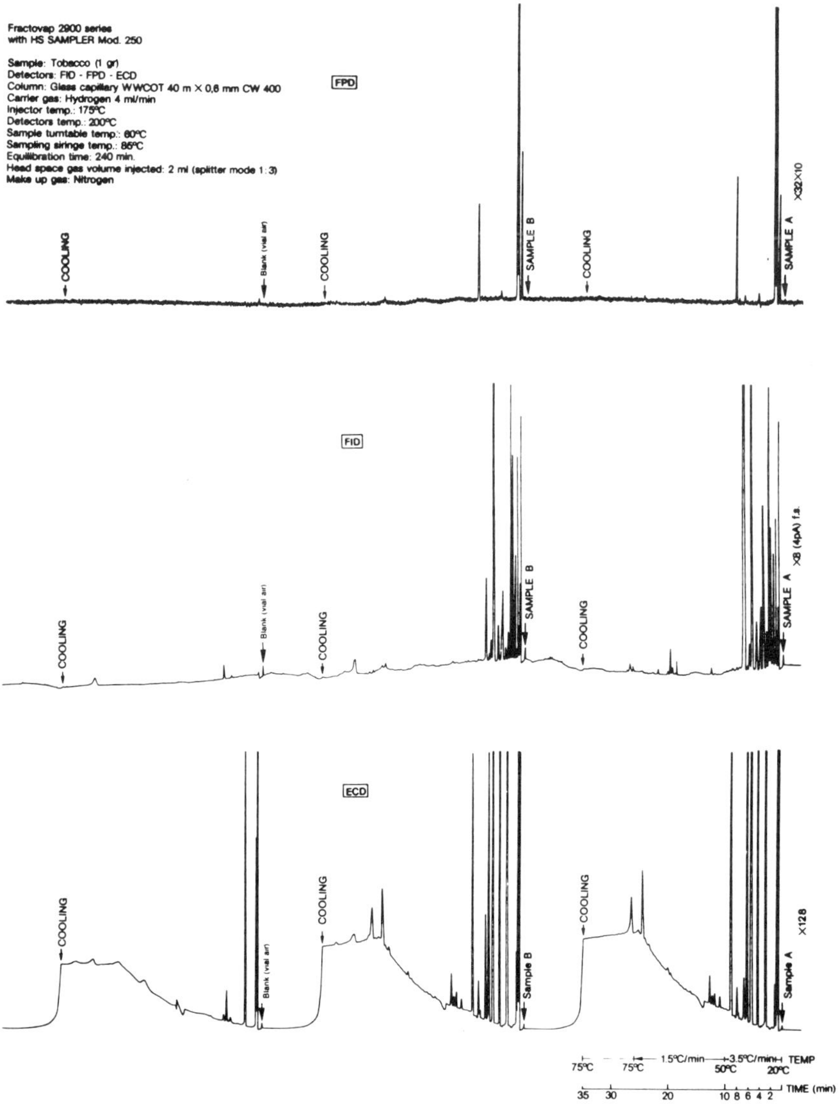

Fig. 11. Head space analysis of tobacco samples with FID-FPD-ECD multidetector system.

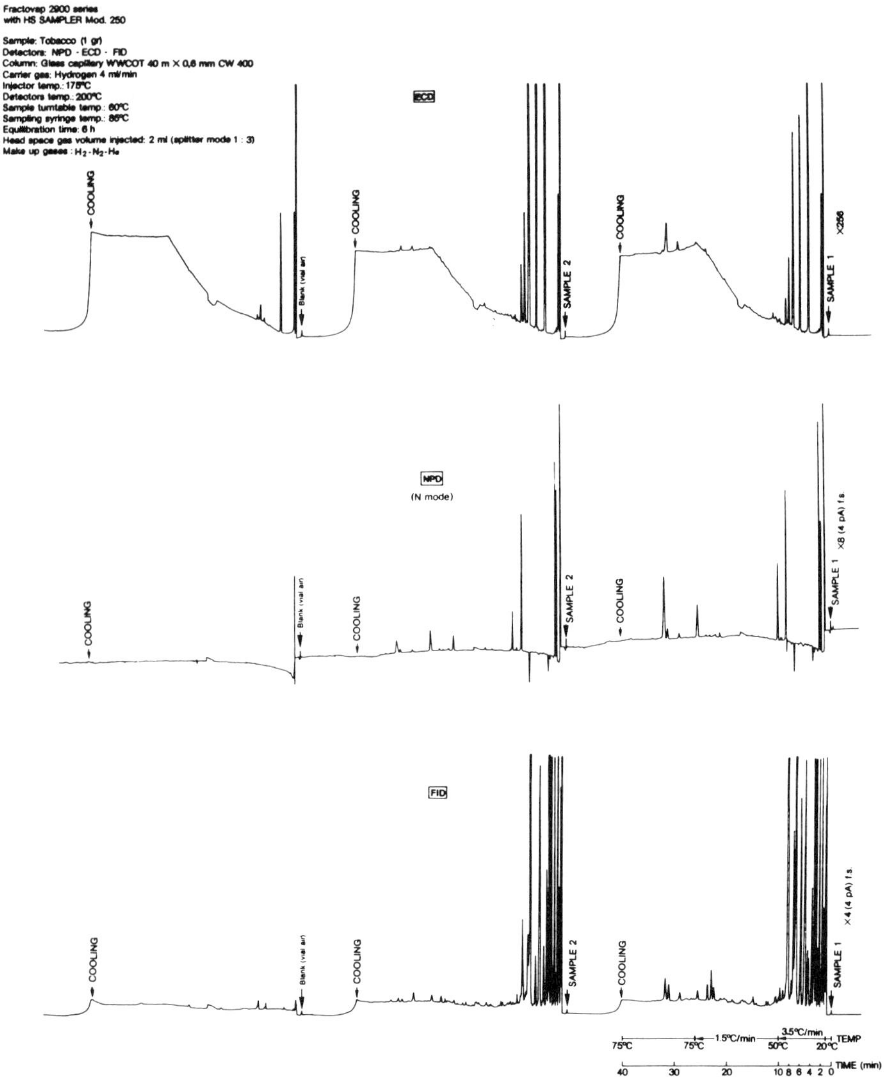

Fig. 12. Head space analysis of tobacco samples with NPD-ECD-FID multidetector system.

Fractovap 2900 series
with HS SAMPLER Mod. 250

Sample: Migration liquid (olive oil from ABS resin)
Detectors: FID - NPD
Column: Glass capillary WCOT 25 m x 0.34 mm
(JXR film thickness 1.4 μm)
Carrier gas: Hydrogen 2.5 ml/min
Column temp.: 40 °C
Injector temp.: 200 °C
Detectors temp.: 200 °C
Sample turntable temp.: 80 °C
Sampling syringe temp.: 90 °C
Equilibration time: 60 min
Head space gas volume injected: 2 ml (splitter mode 1:4)
Make up gases: N_2 - He

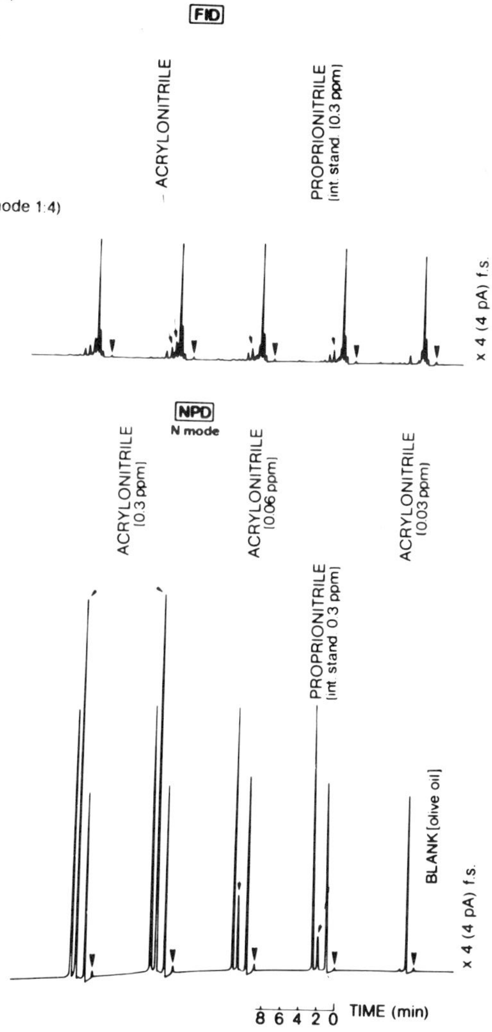

Fig. 13. Acrylonitrile determination in olive oil migration liquid.

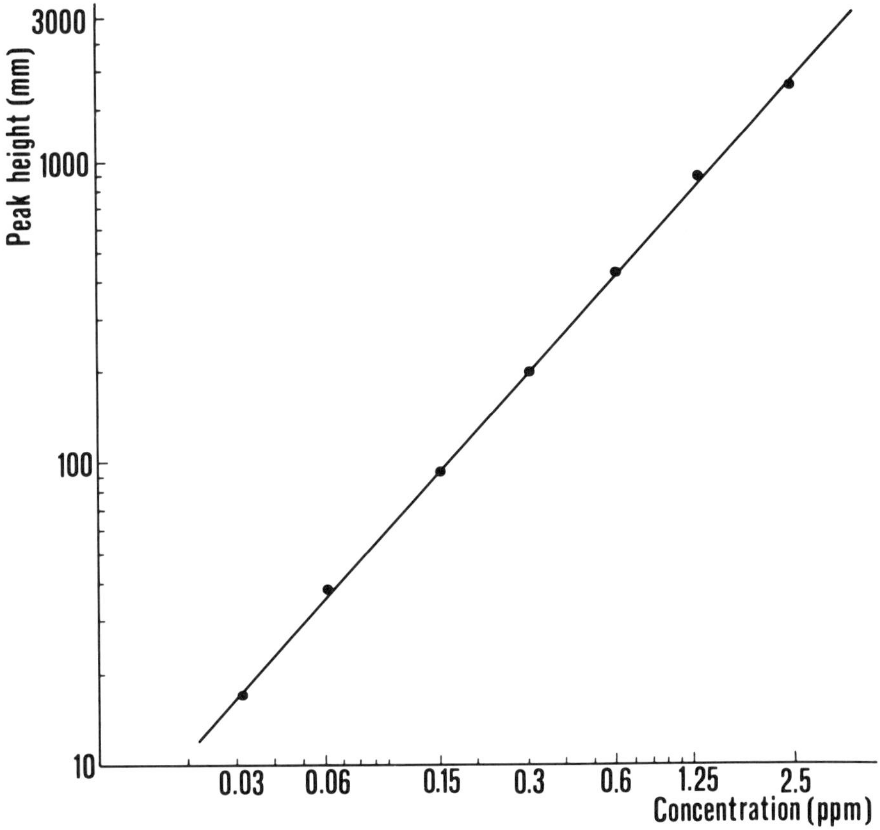

Fig. 14. Calibration curve of acrylonitrile in olive oil with NPD detector.

IV. CONCLUSIONS

Direct analysis of head space gases using high resolution capillary columns may be very useful in determining volatile compounds present as traces in complex mixtures,such as aromas, and in quality control.

The use of multidetector configurations gives a large amount of information in a much shorter time than would otherwise be required. In addition, the possibility of fully automatic HS analysis means that a large number of analyses can be completed with excellent reproducibility of quantitative data.

ACKNOWLEDGMENTS

We thank Mr. A.Sironi for his assistance in the preparation of the paper and Miss. O. Sestagalli for typing the manuscript.

REFERENCES

1. Grob, K., and Grob, G., Proceedings of the 5th International Tobacco Scientific Congress, Hamburg, 1970.
2. Nurok, D., Anderson, J.W., and Zlatkis, A., Chromatographia 11 (4) (April 1978).
3. Analysis of Foods and Beverages Head Space Techniques, edited by George Charalambous (Academic Press, New York, 1978).
4. Brown, R.H., and Purnell, C.J., J. Chromatogr. 178, 79-90 (1979).

5. Simpson, R.F., Chromatographia 12 (11), 733 (November 1979).
6. Drozd, J., Novakand, J., and Rijks, J.A., J. Chromatogr. 158, 471 (1978).
7. Schomburg, G., Husmann, H., and Weeke, F., J. Chromatogr. 112, 205-217 (1975).
8. Johansen, N.G., Chromatogr. Newslett. 7, 18-21 (1979).
9. Sandra, P., Saeed, T., Redant, G., Godefroot, M., Verstappe, M., and Verzele, M., J. High Res. Chromatogr. 3, 107-114 (March 1980).
10. Sandra, P., Vanluchene, E., and Verzele, M., Chromatographia 8, 499 (1975).
11. Mackay, D.A.M. and Hussein, M.M., J. Chromatogr. 176, 291-303 (1979).
12. Parliment, T.H., Marcia D. Spencer, paper 482 Pittsburgh Conference 1981.
13. Hrivnač, M., Frischknecht, W., and Cechova, L., Anal. Chem. vol. 48, No. 6, May 1976 937-940.
14. Gabri, G., and Chialva, F., J. High Res. Chromatogr., Vol. 4, May 1981, 215-217.
15. White, F.H. and Wainwright, T., J. Inst. Brew. 81, 37 (1975).
16. Sandra, P., Cerevisiae 1, (1976).
17. Niefind, H.J. and Spaeth, G., Proc. Am. Soc. Brew. Chem. 54 (1975).
18. Slotema, F.P., Brewers Dig. 53, 48-52 (1978).
19. Gagliardi, P., Verga, G.R., Munari, F., Amer. Lab. May 1981, 82-92.
20. Di Pasquale, G., Di Iorio, G., Capaccioli, T., Gagliardi, P., Verga, G.R., J. Chromatogr. 160 (1978) 133-140.

APPLICATION OF FUSED SILICA CAPILLARY COLUMNS FOR FLAVOR ANALYSIS

Takayuki Shibamoto

Department of Environmental Toxicology
University of California, Davis
Davis, California

I. INTRODUCTION

Impressive advances in flavor analysis began with the invention of the gas chromatograph, a sophisticated analytical instrument based on the principle of the separation of mixed components. In 1952, James and Martin invented this instrument and developed the quantitative analysis of mixtures (1), another important function of gas chromatography. One of the most striking features of gas chromatography is that qualitative and quantitative analysis can be performed simultaneously. This is of incalculable use to flavor chemistry, as flavors depend not only on the constituents of a substance but also on their relative quantities.

The main components of flavors obtained from various food samples are low boiling point organic compounds called volatiles. An important parameter for the separation of components in gas chromatography is partitioning between the gas and liquid phase. Substances which are readily vaporized such as volatiles are, therefore, ideal for gas chromatographic analysis. There are several reference books available describing the principles of gas chromatography (2,3,4,5).

CHEMISTRY OF FOODS AND BEVERAGES:
RECENT DEVELOPMENTS

ISBN 0-12-169080-6

II. COLUMN DEVELOPMENT

The most important part of a gas chromatograph is the column, as this where separations are performed. The first column used was a packed column, the resolution of which was poor compared with current capillary columns.

In 1952, Griffiths *et al.* had already attempted to separate volatile organic mixtures (benzene, toluene, o-xylene, chlorobenzene, etc.) using a charcoal column (6). Subsequently, James and Martin introduced a gas-liquid partition column. They succeeded in separating a series of fatty acids (acetic - n-valeric) using an 11-foot stainless steel packed column (liquid phase = DC 550 silicone). In this report they defined the following two properties, which have proved to be the most important principles of gas chromatography (7).

1. The partition coefficient between the two phases is a constant.
2. Diffusion within a single phase along the length of the column is negligible.

They also succeeded in separating several other organic compounds: aliphatic amines, pyridines, and higher fatty acids (8,9). Since the pioneer work of James and Martin, many researchers have worked to improve column resolution. Figure 1 shows a gas chromatogram of peppermint oil obtained by a classic packed column.

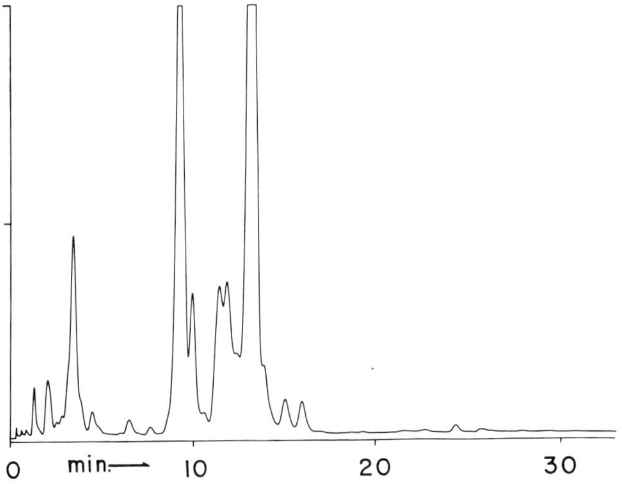

FIGURE 1. Peppermint oil on 6m x *6 mm i.d. (CW 20M).*

The resolution obtained by a packed column is less than 20,000 theoretical plates.

In 1957, Dr. M.J.E. Golay invented a revolutionary column called the open tubular column (10). They are also called capillary or Golay columns. There are three varieties of this column: wall coated open tubular (WCOT), porous layer open tubular (PLOT), and support coated open tubular (SCOT). WCOT columns are the most commonly used. This type of column can give 100,000 - 200,000 theoretical plates.

The invention of a capillary column permitted significant progress in flavor analysis. Figure 2 shows a gas chromatogram of the same peppermint oil as Figure 1 obtained by an early type of stainless steel capillary column.

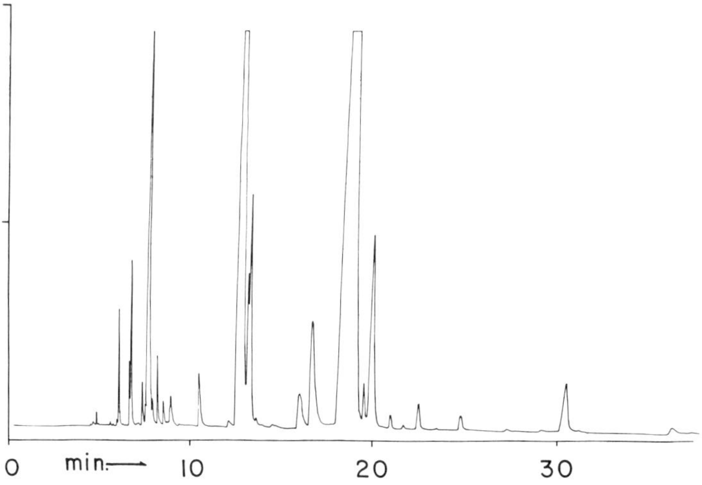

FIGURE 2. Chromatogram of peppermint oil: 45 m X 0.5 mm i.d. stainless steel open tubular column coated with Carbowax 20M and programmed from 100 to 175°C at 3°C/min; nitrogen carrier gas flow rate 15 cm/sec.

One annoying problem caused by a stainless steel column, however, is that the active metal surface promotes some chemical changes in the sample components. This is especially true for an essential oil analysis because such samples contain many terpenes which are extremely susceptible to interconversion reactions. These reactions are encouraged by the metallic surfaces of stainless steel columns (11,12). To solve this problem, glass columns have been used ever since the beginning of column development.

The method of drawing glass capillary tubing was invented

in 1960 (13) and many scientists have worked to prepare high resolution glass capillary columns. In the early 1960's, there were problems preparing a stable liquid phase coating on the glass surface. The polar liquid phases such as Carbowax 20M seemed to have more trouble than non-polar ones for coating. As a result of intensive work on coating techniques by many researchers, today stable glass capillary columns with high resolution are commercially available. Figure 3 shows the gas chromatogram of the same oil as Figures 1 and 2 obtained by a recent glass capillary column.

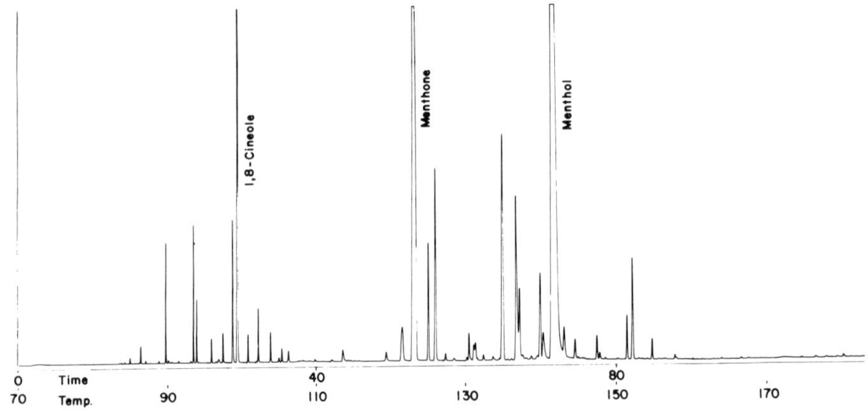

FIGURE 3. Chromatogram of peppermint oil: 100 m X 0.28 mm i.d. glass capillary column coated with Carbowax 20M and programmed from 70 to 200°C at 2°/min; nitrogen carrier gas flow rate 15 cm/sec.

Excellent reviews of the glass capillary column and its applications are available in the literature (13,14). There are, however, still some problems in the use of glass capillary columns:

1. Inside coating is somewhat difficult for private laboratories.
2. Handling and installing of the columns are difficult and require some training because of the column's fragility.
3. The column surface is not completely inert, a fact which causes tailing on polar samples such as acids, polyhydroxy compounds etc.

Fused silica capillary columns appear to have solved some

of the above problems. Of course, even the fused silica capillary column has some problems:

1. They are difficult to manufacture in private labs.
2. One must coat the outside surface with polyimide to reduce fragility.
3. A special column rack is required.

Figure 4 shows a gas chromatogram of the same oil as Figures 1, 2, and 3 obtained with a fused silica capillary column.

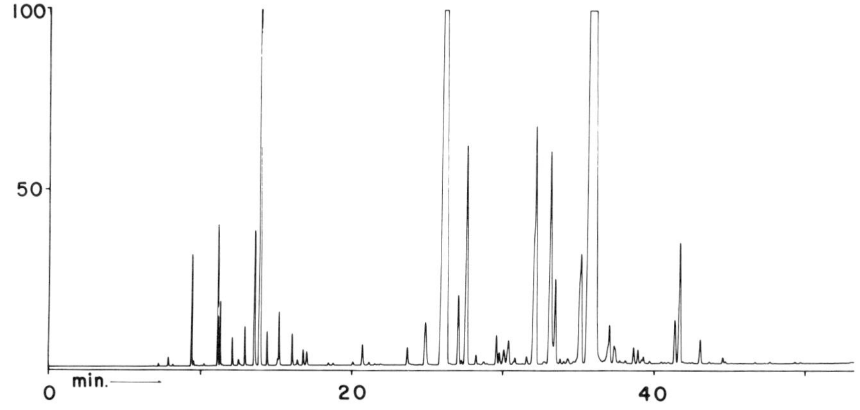

FIGURE 4. Chromatogram of peppermint oil: 50 m X 0.23 i.d. fused silica capillary column coated with Carbowax 20M and programmed from 80 to 220°C at 2°C/min; nitrogen carrier gas flow rate 15 cm/sec.

III. FUSED SILICA CAPLILLARY COLUMNS

A. *Literature Review*

The history of the fused silica capillary column is very short. Dandeneau and Zerenner reported applications of fused silica capillary columns for the first time in 1979 (15). They reported that fused silica proved to be an ideal material for capillary column construction, being inherently more inert than glass containing metal oxides. The high flexibility of fused silica columns also solved many mechanical problems associated with glass capillary columns, such as fragility and column straightening. Further, fused silica columns coated with SP-2100 (deactivated only once with Carbowax 20M) showed tail free peaks on alcohols, phenols, amines, and even on free fatty acids.

Lipsky *et al.*compared two types of columns, natural quartz and fused silica, and obtained superior results most often from fused silica column. Their chromatograms showed that fatty acids do not tail on a fused silica column but do on a natural quartz column, when both are coated with Carbowax 20M (16). The superiority of fused silica was also demonstrated by the results obtained from volatile metal chelate compounds (17). The positinal isomers of biological important alkene halohydrins (1-bromo-2-propanol, 2-bromo-1-propanol, 1-iodo-2-propanol, 2-iodo-1-propanol) were separated completely using a 30 m fused silica capillary column coated with SE-54 (18). The base line resolution of these positional isomers was not obtained by any but fused silica capillary column.

There are only a few reports on food or beverage analysis using a fused silica capillary column. Browning reaction products were analyzed using a fused silica capillary column (20 m x 0.25 mm i.d., OV-101) and methylated polyhydroxy compounds [e.g., 2-(1,2,3,4,5-pentahydroxypentyl)thiazolidine] were chromatogrammed satisfactorily (19). The most recent review on the fused silica column sumarizes the nature of various glass materials, mechanisms of dehydration during column drawing, column injection systems, and several examples of actual applications (20).

B. Column Manufacturing

1. Column Drawing. The fused silica capillary column construction method is not yet widely known and has not been reported in the literature in any detail. We have, however, succeeded in making our own columns, utilizing the schematic diagram of the apparatus shown by Lipsky *et al.*, the only information previously available. Figure 5 shows the drawing and the outside coating instrument prepared by us. More detail is shown in Figure 6. Double coating on the outside tubing significantly increased flexibility of columns. Dupont polyimide pyraline 2525 (solvent: N-methylpyrrolidone) was used for the outside coating.

2. Liquid Phase Coating. We used the static coating method developed for glass capillary columns. Deactivation of the inside surface with Carbowax 20M was also applied prior to coating. Figure 7 shows the apparatus prepared for column coating. The heating zone can be straight, which made the

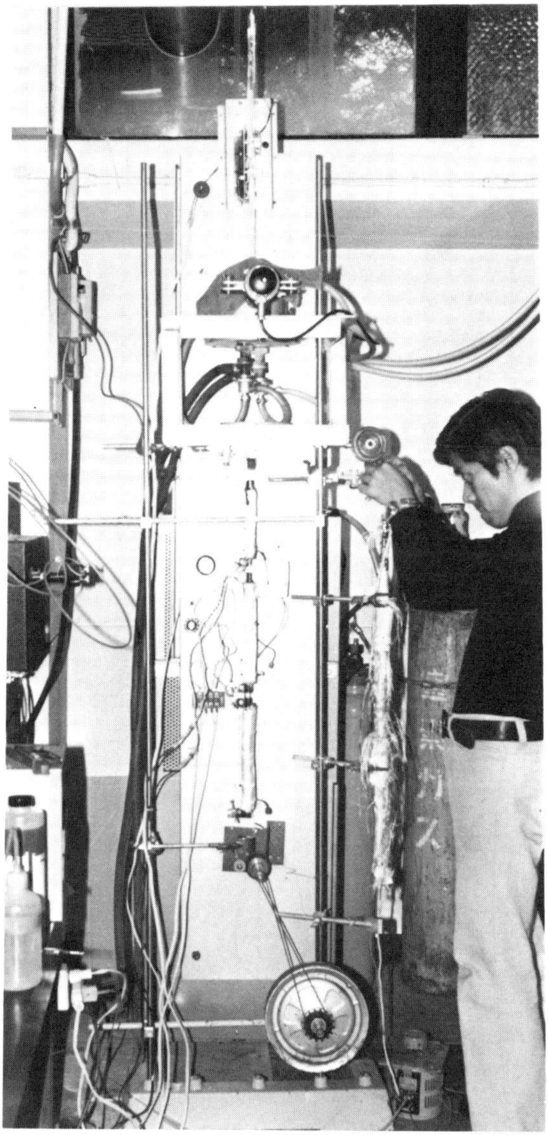

FIGURE 5. Photograph of the drawing and the outside coating instrument.

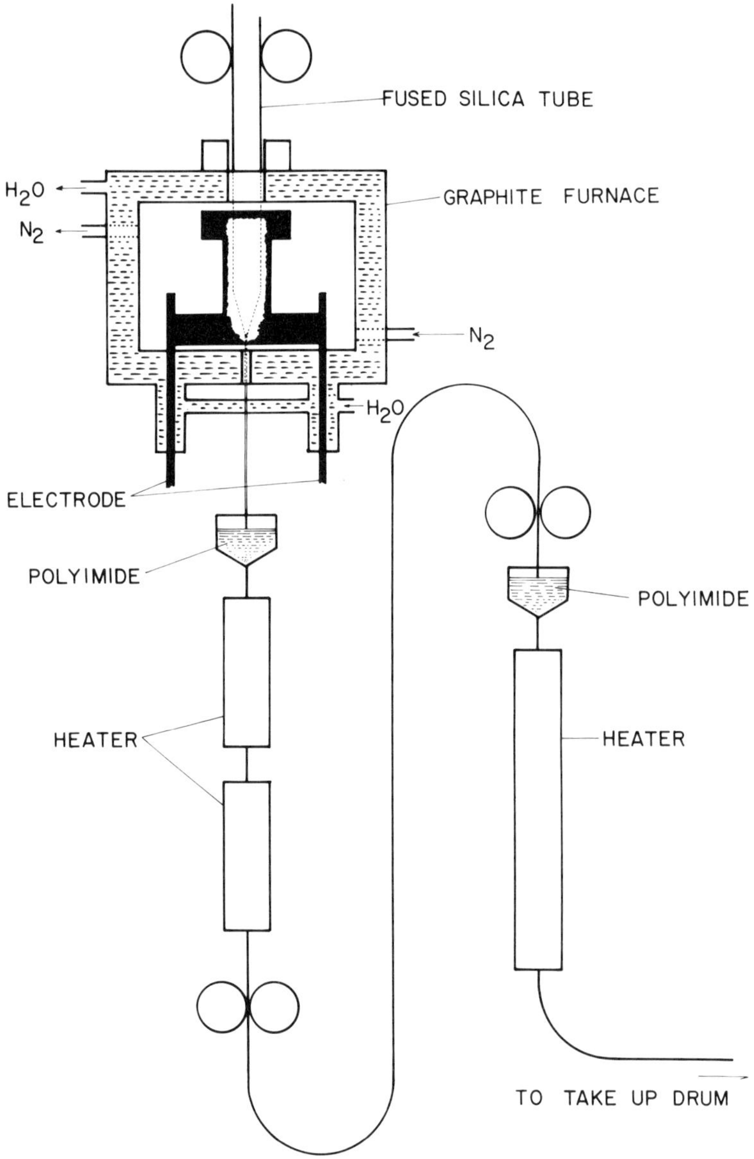

FIGURE 6. Schematic diagram of the drawing and the outside coating instrument.

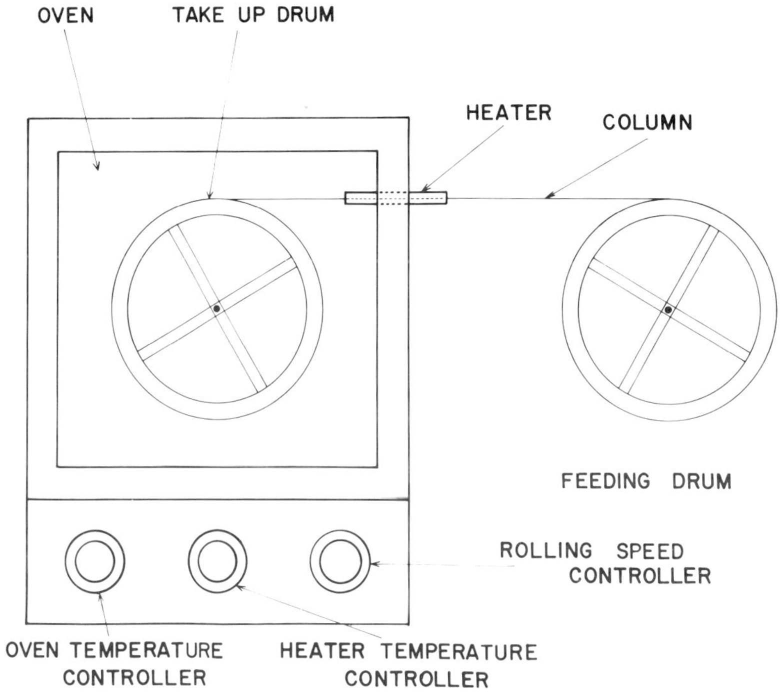

FIGURE 7. Schematic diagram of the liquid phase coating apparatus.

system very simple in comparison with the one for glass capillary columns. All the information in the following sections were obtained using columns prepared in our laboratory. Our columns did not show any inferiority to commercial columns.

C. Inlet Systems

We have developed an all glass inlet system, shown in Figure 8. On-column injection is not used in our experiments because samples of flavor are easy to obtain in sufficient quantities for instrumental analysis. Also, we have not yet obtained satisfactory results from on-column injection. Because of fused silica capillary column's flexibility, two columns can be installed into the injection port and attached to the separate detectors. Simultaneous analysis on two stationary phases (e.g., OV-101 and Carbowax 20M) can be achieved

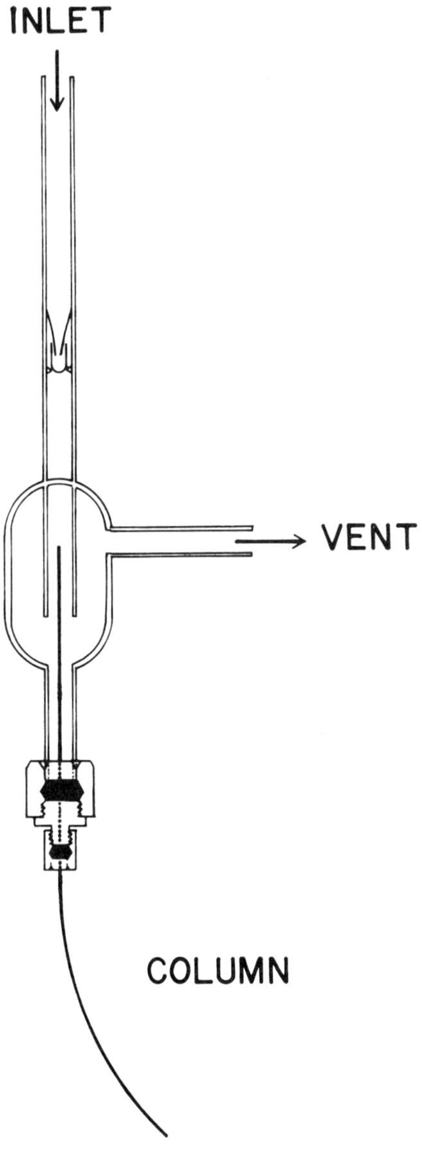

FIGURE 8. All glass inlet system.

by using two fused silica capillary columns. This method offers the ability to maintain identical conditions for two different columns throughout the operation (21). In this method, the end of the two columns can be installed into two different kinds of detector, which eliminates the step of splitting the eluents at the end of a column. Figure 9 shows chromatograms of a mixture (1: thiophene, 2: pyridine, 3: ethylpyrazine) obtained using FID and NPD.

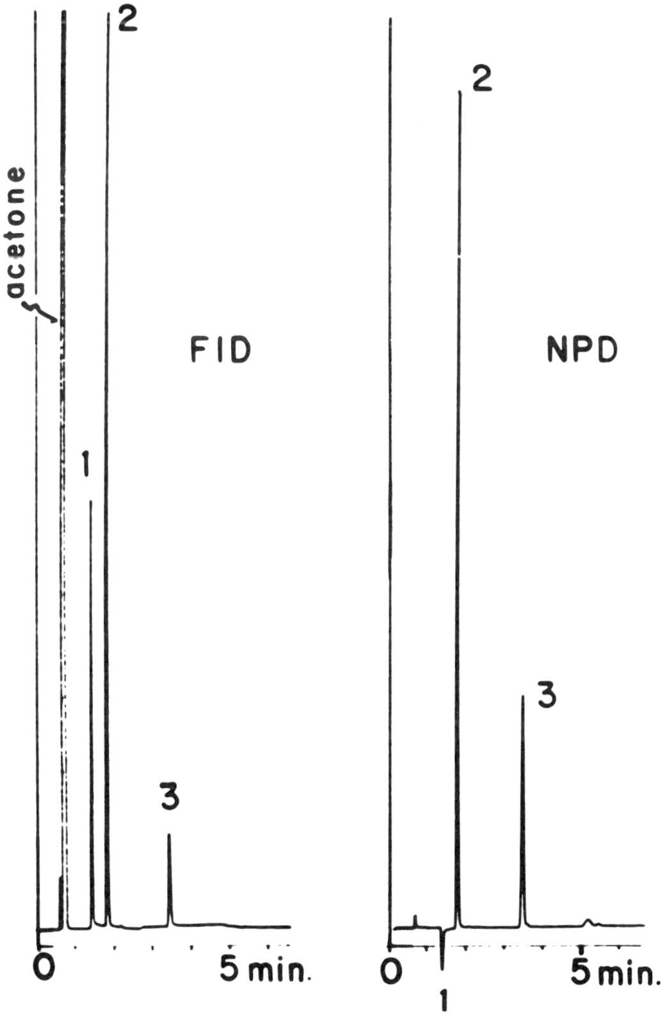

FIGURE 9. Chromatograms of a standard mixture obtained by a dual column system.

We modified our own inlet system for dual channel analysis. Figures 10 and 11 show a modified injection port for dual channel analysis. In order to investigate the difference between glass and fused silica capillary columns, a glass capillary column was installed in the straight part of the splitter outlet and a fused silica capillary column was installed in the branched part of the splitter outlet. The results shown in the following section were obtained using this system.

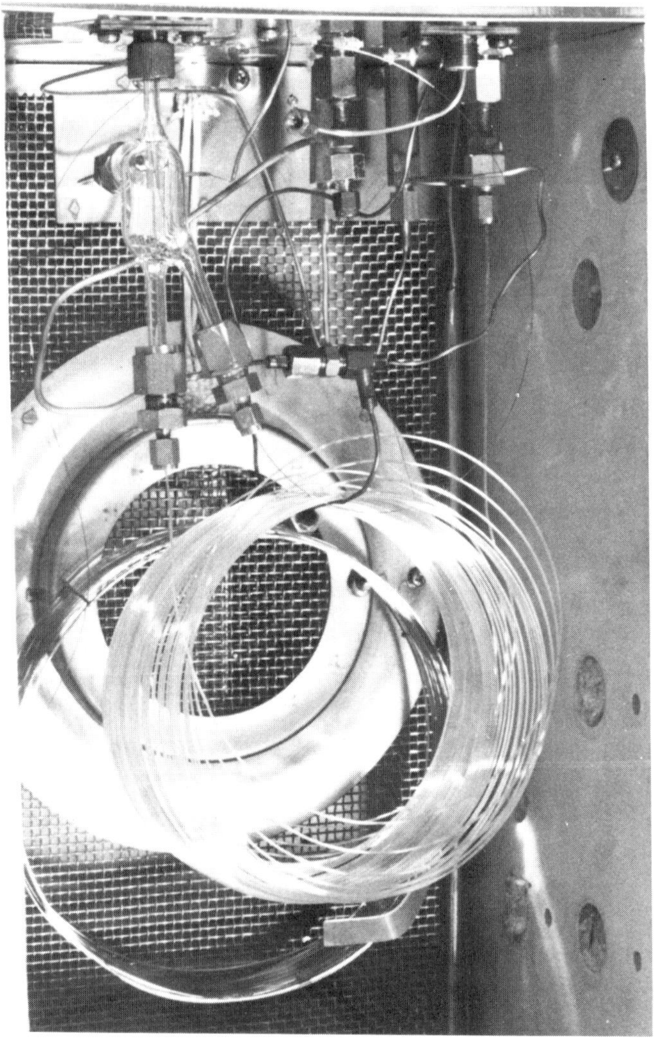

FIGURE 10. Photograph of the dual channel system.

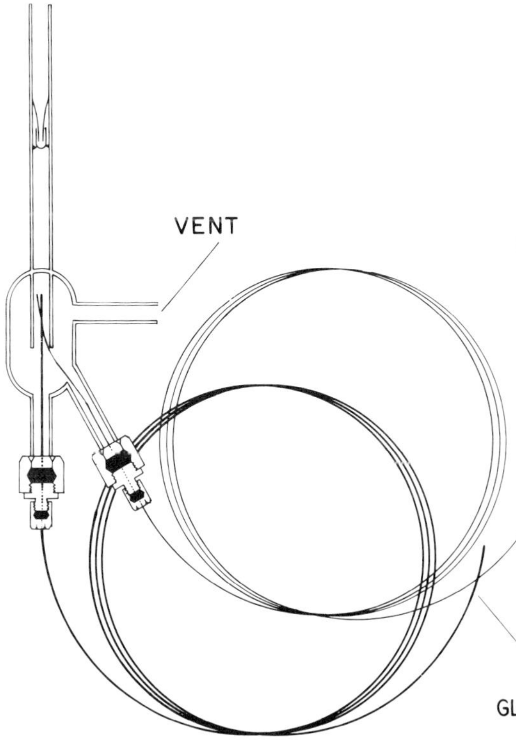

FIGURE 11. Schematic diagram of the dual channel system.

D. Comparison Between a Fused Silica Capillary Column And a Glass Capillary Column

As shown in Figures 3 and 4, resolution is equal in the two columns. A glass capillary column is an excellent column for resolution. The most apparent difference between it and fused silica column is that fused silica gives tail free peaks for polar samples. This is very important in obtaining the high reproducibility in quantitative analyses. The precise results of peak integration can be obtained only from base line resolved peaks. Tailing is, therefore, one of the most undesirable factors in a gas chromatographic analysis. Figure 12 shows gas chromatograms of fatty acid homologues on a glass capillary column and on a fused silica capillary column, both coated with Carbowax 20M. It is obvious that fused silica gives more tail free peaks.

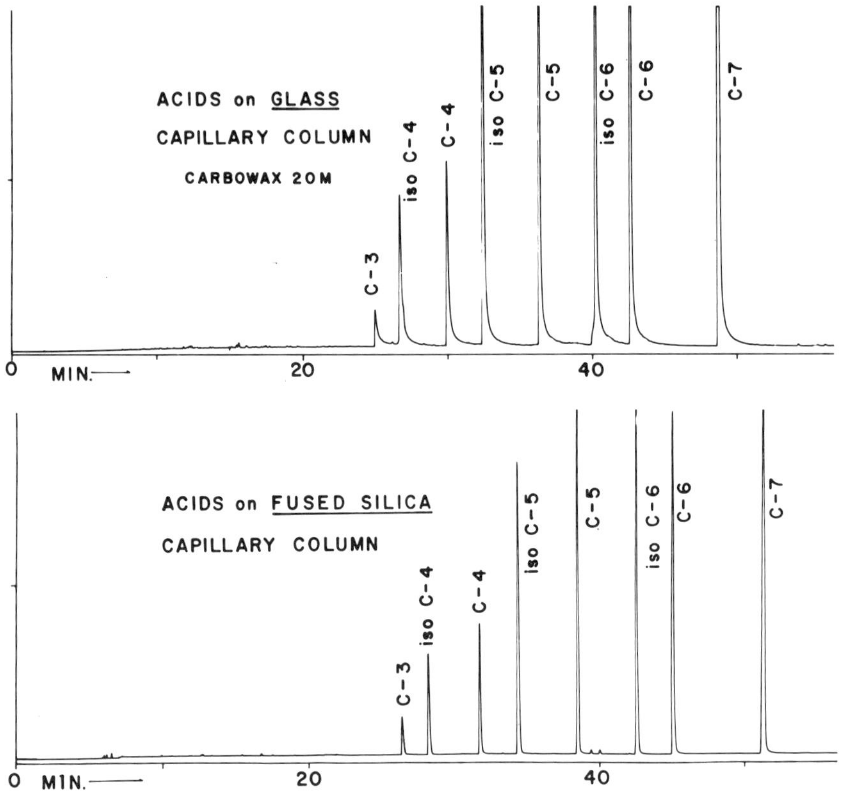

FIGURE 12. Gas chromatograms of a standard fatty acids mixture on a glass capillary column (top, 50 m X 0.28 mm i.d.) and on a fused silica capillary column (bottom, 50 m X 0.23 mm i.d.), bothe coated with Carbowax 20M.

The advantages of having tail free peaks for acids can be seen in the analysis of volatiles samples. Figures 13 and 14 are gas chromatograms of California hop oil on a glass capillary column and on a fused silica capillary column, respectively. Some new peaks are seen in Figure 14 (near the peak of isovaleric acid) due to the removal of tailing from the acid peak.

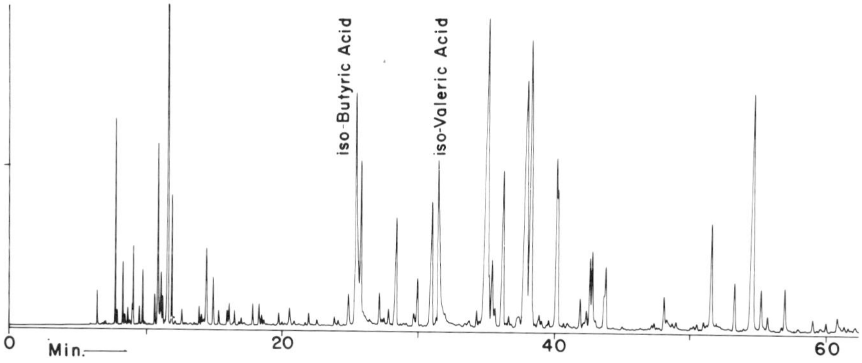

FIGURE 13. Gas chromatogram of California hop oil: 50 m X 0.28 i.d. glass capillary column coated with Carbowax 20M and programmed from 80 to 220°C at 2°C/min; nitrogen carrier gas flow rate 15 cm/sec.

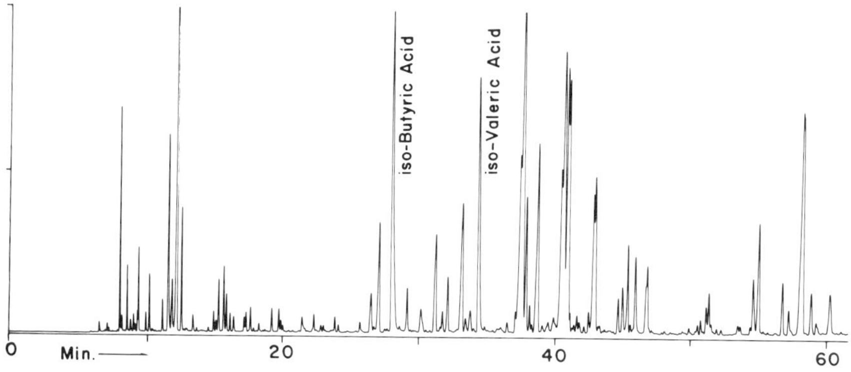

FIGURE 14. Gas chromatogram of California hop oil: 50 m X 0.23 mm i.d. fused silica capillary column coated with Carbowax 20M and programmed from 80 to 220°C at 2°C/min; nitrogen carrier gas flow rate 15 cm/sec.

It must be noted, however, that in our series of routine flavor analyses, we found some contradictory problems in the use of a fused silica capillary column. The acid peaks are easily recognized on a chromatogram from a glass capillary column because of their tailing peaks. This fact makes GC/MS analysis rather easier. Some acids give identical MS fragmentations to their ester derivatives which give no tailing peaks on a chromatogram obtained by a glass capillary column. The acids and their ester derivatives can, therefore, be distinguished using a tailing phenomenon on a chromatogram obtained by a glass capillary column. Figure 15 shows chromatograms of artificial strawberry flavor obtained using a glass capillary column and a fused silica capillary column. The letters a, b, c, d, e, and f indicate the piaks of the component fatty acids.

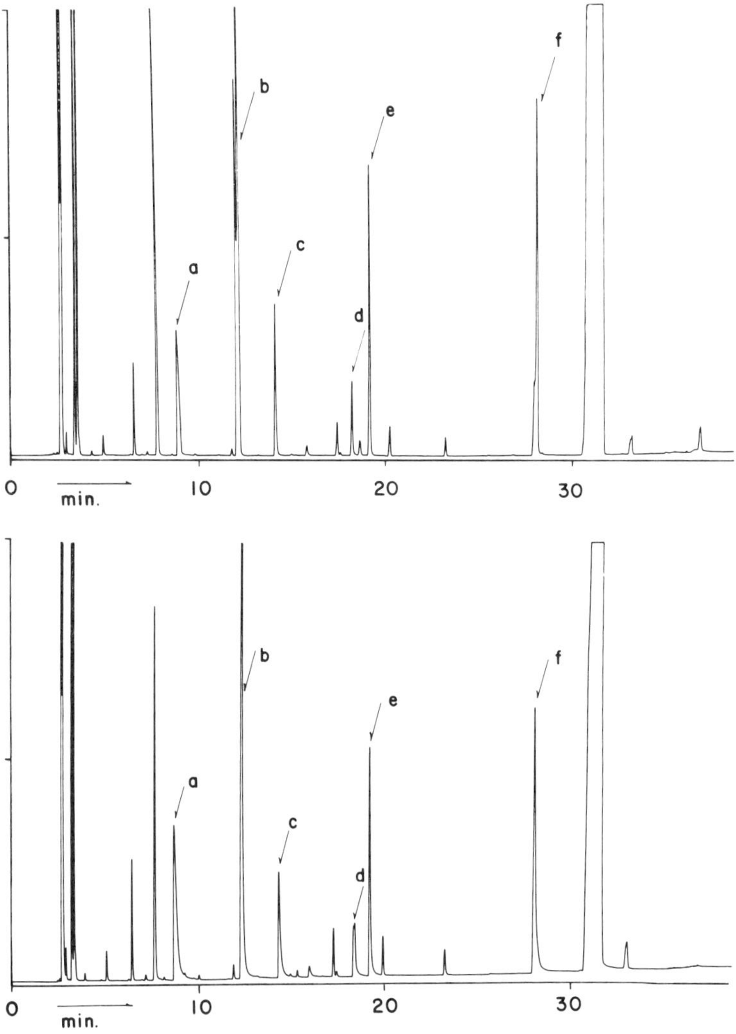

FIGURE 15. Chromatograms of artificial strowberry flavor. Top: fused silica column, bottom: glass capillary column.

The chromatogram (bottom) obtained by a glass capillary column shows tailing peaks of fatty acids which appear as tail free peaks on the chromatogram (top) obtained by a fused silica capillary column. The presence of fatty acids in a volatile sample can be investigated using a glass capillary column. It is still useful, therefore to use a glass capillary columns for numerous aspects of flavor analysis.

IV. APPLICATIONS OF FUSED SILICA CAPILLARY COLUMNS FOR FLAVOR ANALYSIS

Since the fused silica capillary column was invented only two years ago, it is still in the stage of being tested for various uses. Trussell *et al.* reported a series of applications of fused silica capillary column on various test mixtures: EPA base/neutral/acid compounds and pesticides (22).

The following series of gas chromatograms are prepared in our laboratory using a fused silica capillary column coated with Carbowax 20M (50 m x 0.23 mm i.d.). The flavor samples were obtained by entraining on a porous polymer, Porapak Q. The gas chromatogram of the head space samples are shown in Figures 16, 17, 18, 19, 20, 21, 23. The flavor extracts were obtained by organic solvent (methylene chloride) extraction (Figures 22, 24).

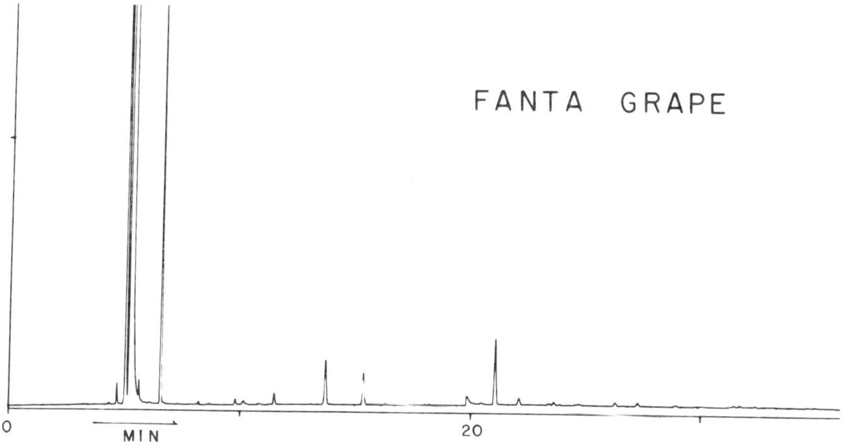

FIGURE 16. A typical gas chromatogram of flavor head space sample, trapped on Porapak Q. Column: 50 m X 0.23 mm i.d., coated with Carbowax 20M.

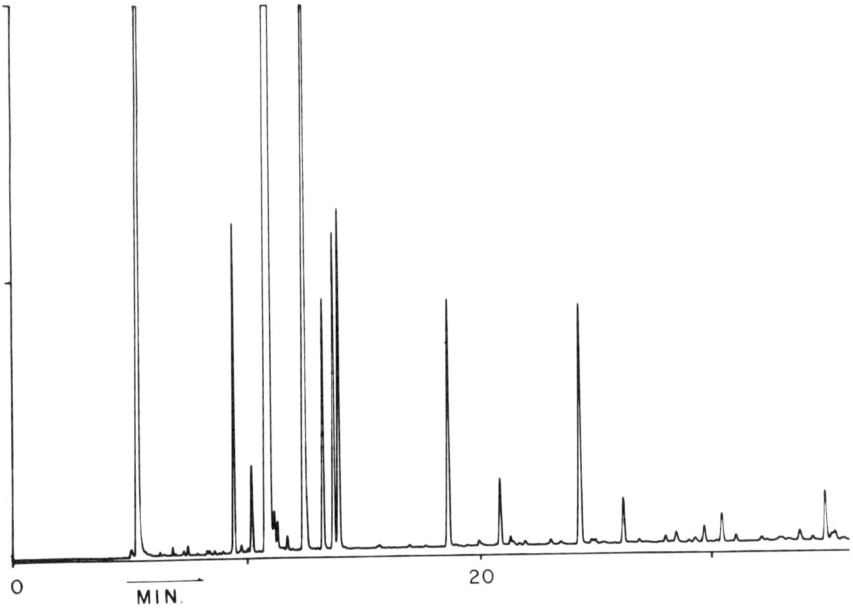

FIGURE 17. Gas chromatogram of head space sample from an American soft drink.

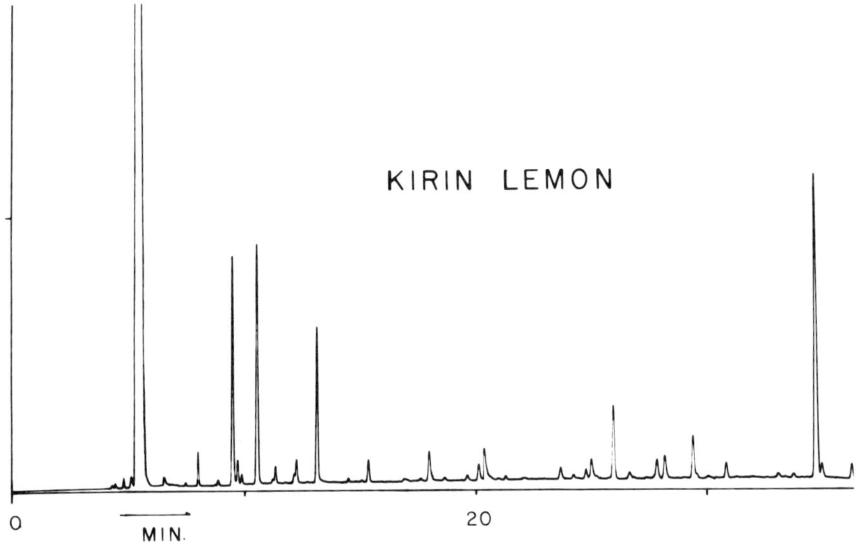

FIGURE 18. Gas chromatogram of head space sample from a Japanese soft drink.

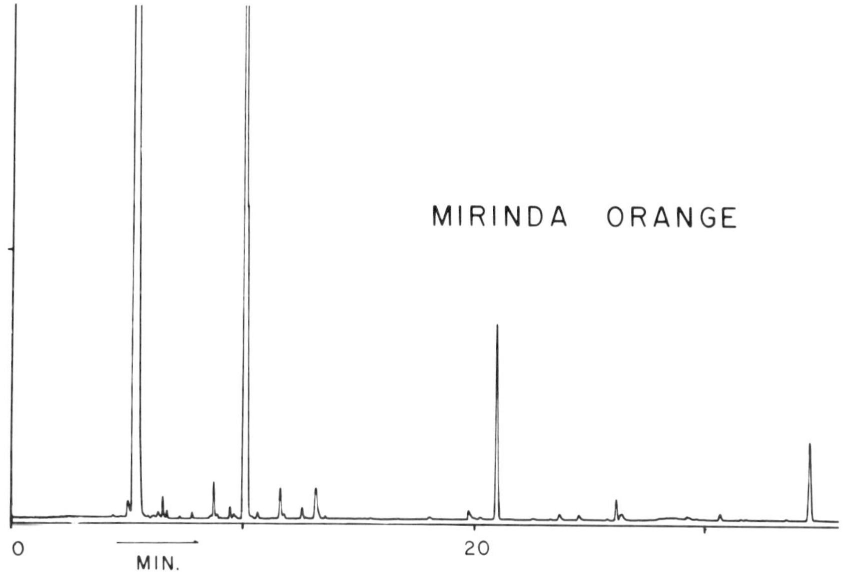

FIGURE 19. Gas chromatogram of head space sample from a citrus drink.

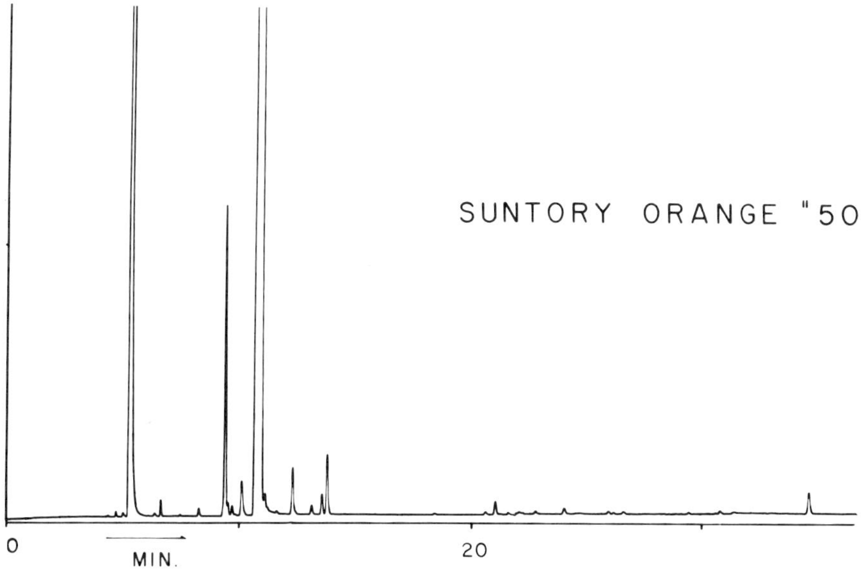

FIGURE 20. Gas chromatogram of head space sample from a citrus drink.

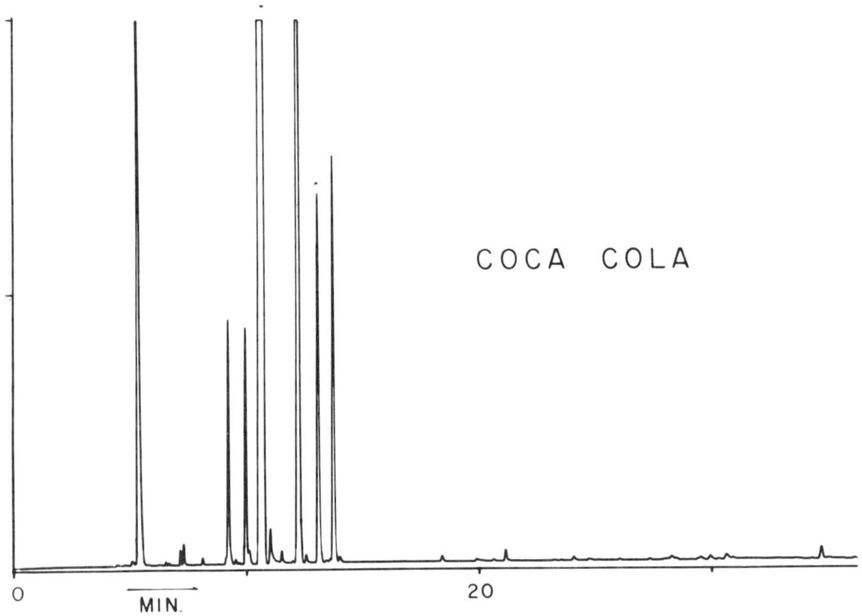

FIGURE 21. Gas chromatogram of head space sample obtained from Coca-Cola.

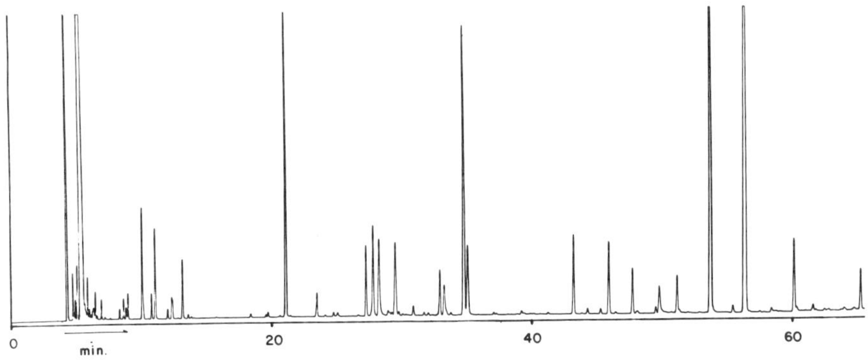

FIGURE 22. Gas chromatogram of methylene chloride extract obtained from Coca-Cola.

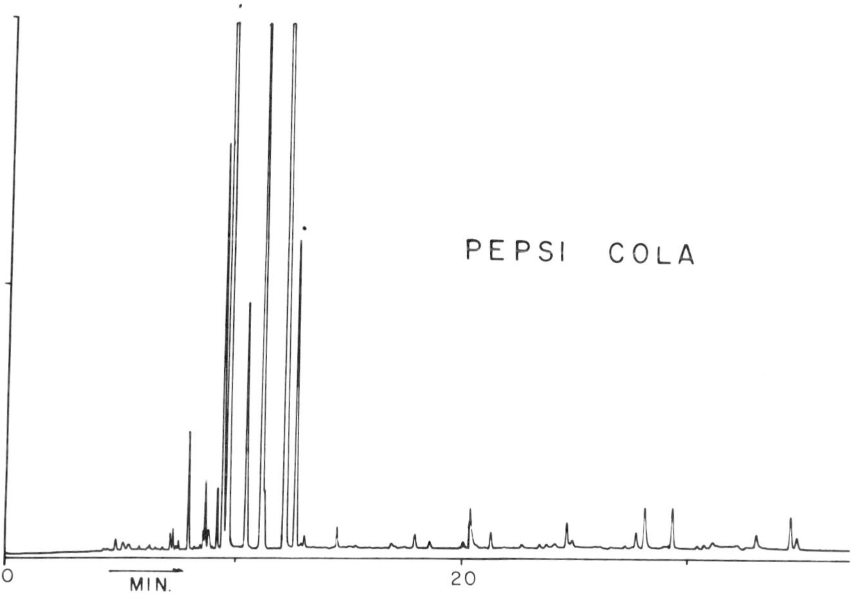

FIGURE 23. Gas chromatogram of head space sample obtained from Pepsi-Cola.

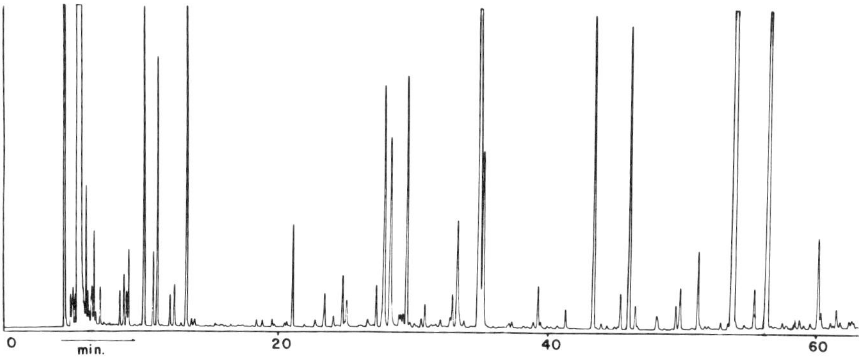

FIGURE 24. Gas chromatogram of methylene chloride extract obtained from Pepsi-Cola.

It is interesting to see the differences between head space samples (Figures 21 and 23) and organic solvent extracts (Figures 22 and 24) obtained from commercial Coca-Cola and Pepsi-Cola.

The samples obtained from a browning reaction mixture generally possess a large number of constituents which vary from polar to non-polar chemicals. A fused silica capillary column gives excellent resolution for these samples. Figure 25 shows a typical chromatogram of a browning reaction sample obtained from a starch/ammonia browning model system by methylene chloride extraction.

V. APPLICATIONS OF FUSED SILICA CAPILLARY COLUMNS FOR FRAGRANCE MATERIALS

Fragrance bases for cosmetic products are constructed from essential oils and synthetic chemicals. The chemicals used for fragrance bases are called aroma chemicals and generally have a low boiling point. Gas chromatography has been applied to investigate aroma chemicals since the invention of the gas chromatograph. Aroma chemistry is, indeed, the field which uses gas chromatography most intensively. There has yet been, however, no report on the application of fused silica column to aroma chemistry.

Some application examples of cosmetic products are shown in Figure 26 (chromatograms of head space and solvent extract samples from a commercial room freshener) and in Figure 27 (chromatogram of a commercial cologne fragrance).

SUMMARY

Fused silica capillary columns have a short history but have already proved their superiority to any existing columns in volatile analyses. Their flexibility has made installation considerably easier and, additionally, provides features such as a dual channel system to gas chromatography. The inertness of the inside surfaces has solved the tailing problem with high polar compounds such as fatty acids and has increased the reproducibility of peak integrations. We predict that fused silica capillary columns will take over the dominant position currently held by glass capillary columns in the near future.

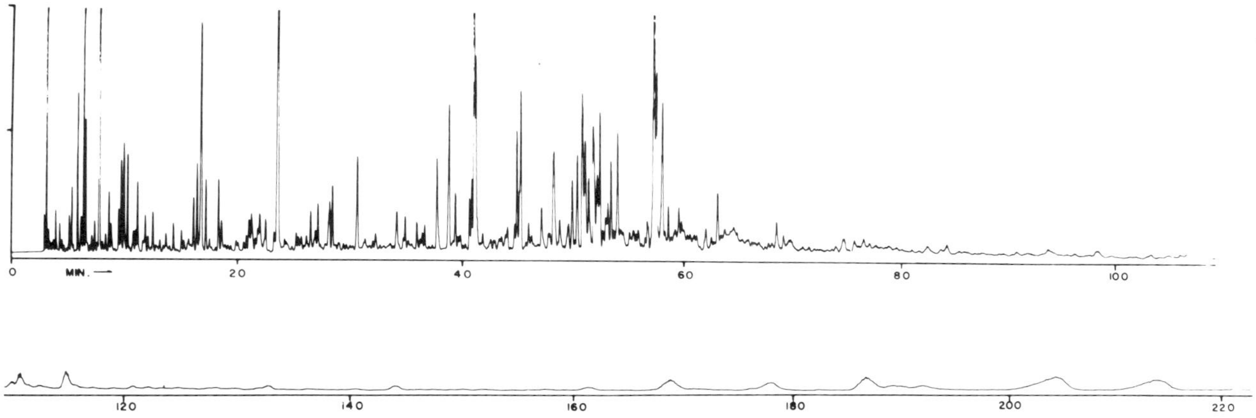

FIGURE 25. Chromatogram of the browning reaction products from a starch/ammonia browning model system.

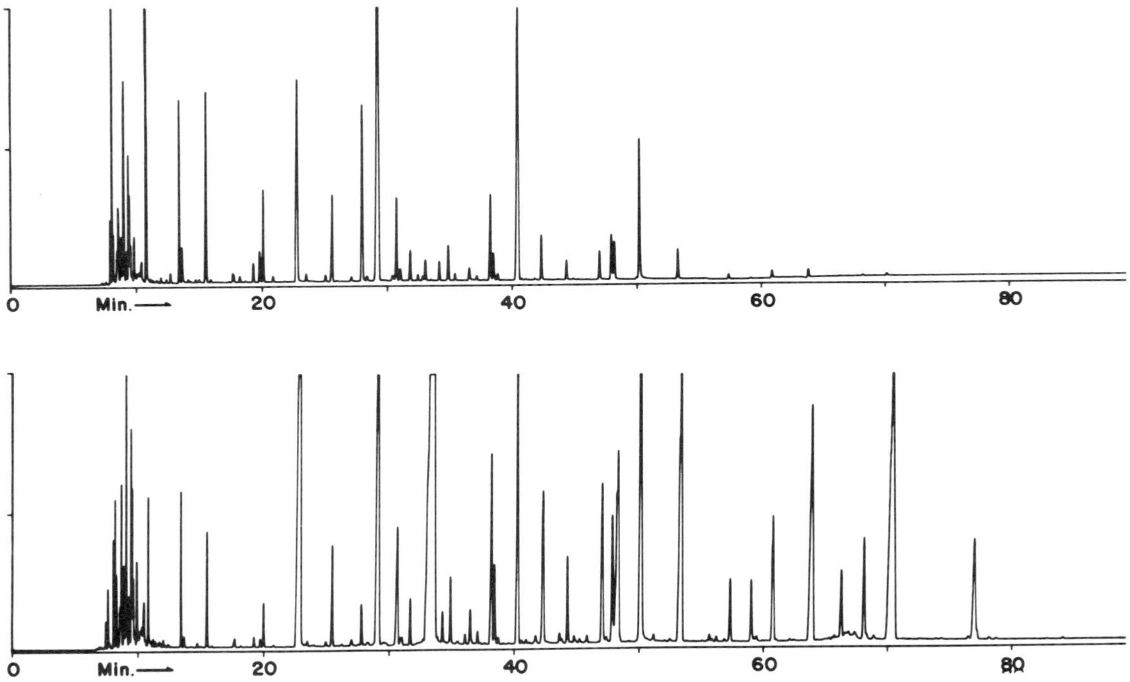

FIGURE 26. Chromatograms of the fragrance samples from a commercial room freshener. Top: head space sample, bottom: pentane extract.

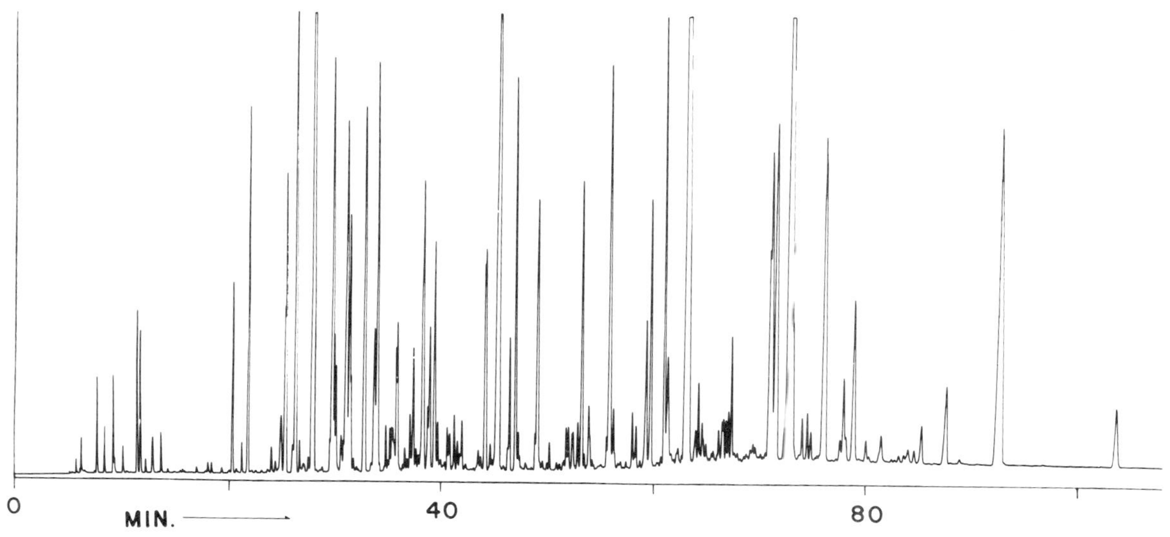

FIGURE 27. Chromatogram of the pentane extract from a commercial cologne.

ACKNOWLEDGMENT

The author gratefully acknowledges K. Yamaguchi, who obtained some of the data for this article, and O. Nishimura, who manufactured the fused silica capillary columns.

REFERENCES

1. James, A. T., and Martin, A. J. P., *Biochem. J. 50*, 679 (1952).
2. Nogare, S. D., and Juvet, R. S. Jr., "Gas-Liquid Chromatography", Wiley (Interscience), New York, (1962).
3. Purnell, H., "Gas Chromatography", Wiley, New York, (1962).
4. Ettre, L. S., and Zlatkis, A. "The Practice of Gas Chromatography", Wiley (Interscience), New York, (1967).
5. Cassidy, H. G., "Fundamentals of Chromatography", Wiley (Interscience), New York, (1957).
6. Griffiths, J., James, D., and Phillips, C., *Analyst 77*, 897 (1952).
7. James, A. T., and Martin, A. J. P., *Analyst 77*, 915 (1952).
8. James, A. T., *Biochem. J. 52*, 242 (1952).
9. James, A. T., amd Martin, A. J. P., *Biochem. J. 50*, 679 (1952).
10. Golay, M. J. E., *in* "Gas Chromatography, Lansing Symposium" (V. J. Coates, H. J. Nobels, and I. S. Fagerson, ed.), pp 1 - 13. Academic Press, New York, (1958).
11. Hunter, G. L. K., and Brogden, W. B. Jr., *J. Org. Chem. 28*, 1679 (1963).
12. Wrolstad, R. E., and Jennings, W. G., *J. Chromatog. 18*, 318 (1965).
13. Jennings, W. G., "Gas Chromatography with Glass Capillary Columns", Academic Press, New York, (1980).
14. Jennings, W. G., and Shibamoto, T., "Qualitative Analysis of Flavor and Fragrance Volatiles by Glass Capillary Gas Chromatography", Academic Press, New York, (1980).
15. Dandeneau, R. D., and Zerenner, E. H., *J. High Resoln. Chromatogr. Chromatogr. Commun. 2*, 351 (1979).
16. Lipsky, S. R., and McMurray, W. J., Hernandez, M., Purcell, J. E., and Billeb, K. A., *J. Chromatog. Sci. 18*, 1 (1980).
17. Sucre, L., and Jennings, W. G., *J. High Resoln. Chromatogr. Chromatogr. Commun. 3*, 452 (1980).

18. Geigert, J., Wadsworth, J., and Moreland, M., *High Resoln. Chromatogr. Chromatogr. Commun. 4*, 184 (1981).
19. Kitamura, K., Wei, C.-I., and Shibamoto, T., *J. Agric. Food Chem. 29*, 378 (1981).
20. Jennings, W. G., *J. High Resoln. Chromatogr. Chromatogr. Commun. 3*, 601 (1980).
21. Phillips, R. J., Wolstromer, R. J., and Freeman, R. R., Hewlett-Packard Application Note AN 228 - 16 (1981).
22. Trussell, A. R., Lieu, F.-Y., and Moncur, J. G., Presented at North American Continent Chemical Congress Symposium on Advances and The Identification and Analysis of Organic Pollutants in Water, Las Vegas, Nevada, August (1980).

A MICRO-OLFACTOMETER FOR CHEMICAL SENSORY ANALYSES

Gerald F. Russell
Judith D. Hills

Department of Food Science & Technology
University of California
Davis, California

INTRODUCTION

The question of how the olfactory system is able to detect and differentiate among odors has been studied for many years. There have been many attempts to build instruments (olfactometers) to meter and control stimuli responsible for olfactory responses. Many refinements in techniques and olfactometer construction have been described in the literature, but often insufficient attention has been given to the purity of the compounds studied. Even minute amounts of impurities in samples may alter significantly the odor qualities of the compounds being tested. With continuing developments in the methods for gas-liquid chromatography (GLC), collection of highly purified volatile compounds is possible. The present study will describe: (1) An olfactometer capable of handling micro-amounts of materials collected from a GLC with capability to quantitate the odor stimuli; and, (2) A sensory evaluation of differences among some highly purified stereoisomers.

ISBN 0-12-169080-6

LITERATURE REVIEW

A wide variety of instruments and techniques developed for determination of olfactory response in humans have been reviewed by Stone et al. (1965) and Amerine et al. (1965). Their development was prompted by the desire to control and measure variables influencing olfactory response, such as (1) concentration and purity of odorant, (2) air flow pressure and volume, (3) air temperature, humidity, and purity, (4) psychological and physiological factors influencing judges.

Although early olfactometers have been superceded by more elaborate devices, no instrument has yet succeeded in perfecting control over all these variables. Most odor evaluation of purified compounds is done by variations of one of three techniques: (1) "blast-injection," (2) sniff-bottles, or (3) air dilution olfactometers, each with its advantages and disadvantages.

The "blast-injection" method, involves direct nasal injection of odorous material under pressure, and has been used mostly in clinical investigations of monorhinal or birhinal stimulation, and in distinguishing between peripheral and central impairment of specific sensory function. This technique is still of doubtful value in psychophysical investigations of olfaction due to difficulties of the judges in differentiating between responses to odor stimuli and responses to the unnatural nasal pressure and environment; it is also probable that the aerodynamics within the nasal passages influences odor detection and discriminaton.

Sniffing from bottles and beakers is the most widely used method because of its simplicity and economy. Difficulty in establishing the vapor concentration above non-ideal solutions as well as a lack of suitable stable non-odorous and non-reactive solvents of low viscosity limits the usefulness of the technique.

Air dilution olfactometry, in which desired stimulus concentrations are attained by diluting saturated vapor, affords opportunity for both highly controlled conditions and use of natural breathing.

The stimulus is delivered to the subject by such means as: nosepieces, funnels, or enclosures around head or body. Since various devices can be used to provide fairly precise control of air flow rates, temperature, humidity and stimulus concentration, this type of olfactometer has been increasingly employed in studies on the psychophysical aspects of olfaction. Such equipment is, however, often expensive to construct and difficult to calibrate. Also, devices of this type may not lend themselves to rapid change of concentrations of test materials and may be difficult to clean.

Most frequently, control and measurement of stimulus concentration is based on knowledge of the vapor pressure of the test material at different temperatures and the flow characteristics of the system (e.g., Ough and Stone, 1961; Eyferth and Krueger, 1969). Alternatively, stimulus concentration may be controlled and determined by variation of temperature at a sparger when vapor pressures at various temperatures as well as flow rates are known (e.g., Bozza et al., 1960). Loss of weight of test sample over a period of time plus knowledge of flow rates is another method which can be used for calculation of stimulus concentration when saturation does not occur (e.g., Stuiver, 1960). GLC quantitation has also been employed for determining concentration of diluted, non-saturated vapors. Such quantitation, was employed in an early study by Johnston (1967); it was not rapid, but sufficed to predetermine the constant levels maintained by sparging of air through the test material.

Although air dilution olfactometers offer advantages over other methods of olfactometry, the designs have often limited investigations to those compounds easily purified by large scale methods such as distillation, since large volumes of test material were required for sparging. Purity is seldom assured, at least at the levels measurable by modern instrumentation which even then may not suffice as a criterion of purity for olfactory measurements. An olfactometer capable of presenting and measuring small amounts of highly purified materials such as obtained from GLC purification or analysis would be of considerable advantage in theoretical studies on odor perception.

We describe in this chapter the design of a simple microscale olfactometer capable of presenting judges with carefully controlled and highly purified chemicals. An example of the use of the instrument is presented from further studies of previously reported researches on the sensory properties of chiral isomers of carvones and related compounds (Russell and Hills, 1969, 1971; Friedman and Miller, 1971).

APPARATUS AND EXPERIMENTAL PROCEDURE

A. Design of the Olfactometer

An olfactometer was designed and constructed that would require comparatively small amounts of sample. This feature made feasible the testing of highly purified chemicals collected from a GLC. The design of the olfactometer, shown schematically in Figure 1a, was based on the principle of air dilution. However, unlike most air dilution olfactometers, the concentration of odorant delivered was measured directly by a GLC rather than derived through calculations based on dilution of a known volume of saturated air.

Nitrogen, the carrier gas for the olfactometer, entered the system at 10 psi (Matheson Model 70-A low pressure regulator) and passed through a U-tube immersed in a dry ice/ethanol bath and filled with 60-80 mesh crushed firebrick. The emergent stream of purified nitrogen was then channeled by two 3-outlet glass manifolds into the main carrier lines or the capillary lines which contained the odorant stimuli.

The nitrogen entering the three capillary lines was metered by fine control stainless steel needle valves. It then passed through the capillary collection tubes which were inserted into the glass system by means of silicone rubber septa. The nitrogen, carrying odorant molecules, flowed into a glass "tee" where it mixed with the larger flow of nitrogen of the main carrier line, also metered by fine control needle valves.

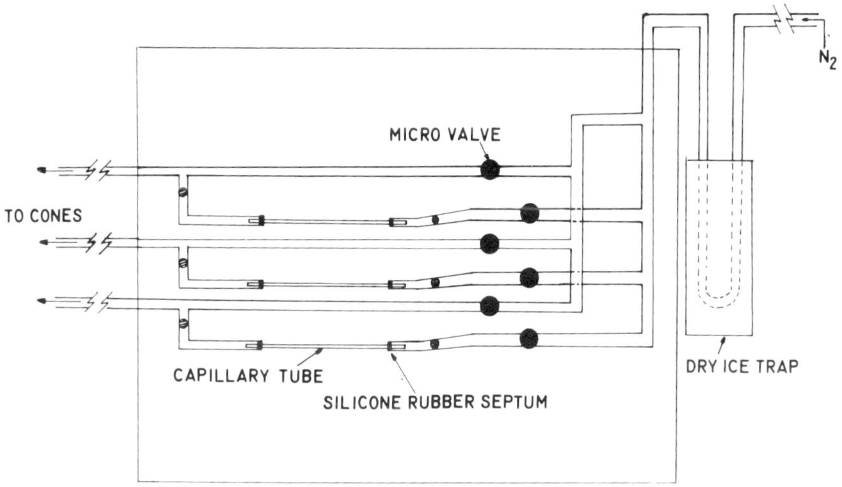

FIGURE 1a. Olfactometer main panel.

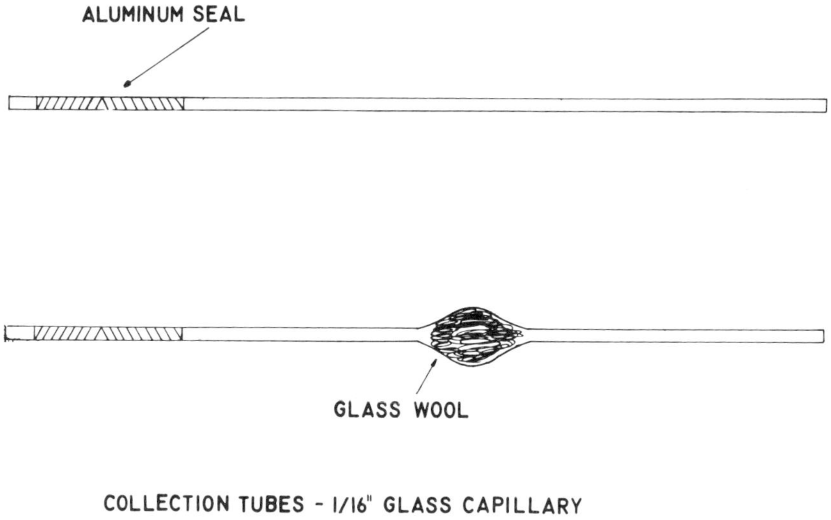

FIGURE 1b. GLC collection tubes for odor stimuli.

The main control panel of the olfactometer, which was mounted on a 0.7 m x 1 m section of Lucite, was constructed of glass tubing, stainless steel valves, Teflon connecting tubing, and silicone rubber septa at the capillary tubes. Odorant stimuli were transported from the main control panel to three Pyrex nose cones for sniffing by lines of 5 mm i.d. Teflon tubing. Pyrex nose cones were mounted at 30 cm intervals at the front of a fume hood. The Teflon tubing at each tee" could be disconnected and a monitoring line to a GLC connected (vide infra).

B. Standardization of Stimulus Concentration

Stimuli concentrations at the "tees" were measured using a Varian model 1200 flame ionization detection (FID) GLC. The odorant gas stream passed through an insulated 5 mm i.d. stainless steel tube maintained at 225 deg C into a Carle sampling valve with two 0.5 mL loops, also maintained at 225 deg C. When the valve was turned, 0.5 mL of carrier gas containing volatiles was injected directly onto a 3 mm i.d. x 1.5 m column of 5% SE 30 on Chromosorb G, maintained isothermally at 220 deg C. Peaks from volatiles were recorded on a Varian model 20 recorder and integrated by an Infotronics CRS-104 integrator. Operating conditions of the GLC and digital integrator are given in Table 1.

Digital response to carvone previously purified by GLC and measured at the olfactometer was standardized and calibrated to ng/mL of odorant in nitrogen by the following procedure. A 2.5 ul aliquot of

TABLE 1. GLC Conditions for Vapor Quantitation

Instrument:	Varian 1200 GLC with FID
Carrier gas:	Nitrogen, 25 mL/min
Injector temp:	240 deg C
Detector temp:	275 deg C
Sampling valve temp:	225 deg C
Column temp:	Isothermal 220 deg C
Column type:	Stainless steel, 3 mm x 1.5 m i.d. packed with 5% SE30 on 70/80 mesh Chromosorb G

purified carvone was weighed accurately and diluted with chromato-quality carbon disulfide to 50 mL in a volumetric flask. Known volume samples of this solution were injected directly onto the column of the FID GLC.

C. Preliminary Studies

1. Collection tubes. Conventional 30 cm melting point capillary tubes were used initially for GLC collection and olfactometer dispensation of chemicals tested on the olfactometer. When GLC collected samples were large enough to occlude the tube (approximately 2 ul) nitrogen flow caused the sample to move gradually along the tube and sometimes out the end into the glass framework of the olfactometer. Sudden concentration changes were thus possible as a result of different surface area of sample exposed to the nitrogen flow.

To overcome this problem, a capillary tube with a bulb in the center was made by heating and drawing out a disposable pipet. However, the bulb alone was not found to act as a satisfactory receptacle for the chemical collected from a GLC since capillary action tended to draw sample into the narrow part of the tube. At high nitrogen flow rates the sample remained stationary just in front of the bulb; at low flow rates, the sample remained stationary further back in the tube. It was observed that at certain intermediate flow rates nitrogen pushed the sample toward the bulb until the force of capillary action caused it to move backward once again. Odorant concentration was found to depend on an oscillation frequency since surface area changes brought about by droplet oscillation precluded accurate regulation by the flow valves as intended, and concentration was not stable and proportional to flow rate.

Introduction of a small amount of silanized glass wood into the disposible pipet before drawing it out corrected these difficulties (see Figure 1b). Collection of odorant compounds from the GLC was made with these tubes in the same manner as with an ordinary capillary tube. Sample which condensed before reaching the bulb was forced into the glass wool by warming the tube over the exit port of the GLC. The sample thus remained stationary once adsorbed onto the glass wool.

2. Stimulus duration. A glass wool collection tube containing 4.8 mg l-carvone was inserted into the olfactometer. Flow rates were adjusted to deliver the carvone, above the recognition threshold, at approximately 240 ng/mL of nitrogen with nitrogen passing through the main carrier line at 100 mL/min and through the capillary line at 20 mL/min. Without altering flow rates, concentration was monitored by GLC every three min until the sample had evaporated.

D. Sensory Evaluation

1. Preliminary preparations. Flow rates through main carrier and capillary lines were measured using a calibrated bubble flowmeter. For the compounds studied in these experiments, the flow through the main carrier lines was maintained at 100 mL/min and the flow through the capillary lines was varied to achieve the desired concentrations. When the capillary tubes were inserted, a cumulative flow rate was measured at the lines marked "to cones" on Figure 1a.

Final collections of odorants were made in the glass wool filled collecton tubes shortly before testing on the olfactometer to avoid the need for flame sealing the tubes and the possibility of pyrolyzing traces of the compound. Final injections varied between 6 and 9 ul, but the amount collected was in the range of 4 to 7 ul, depending on GLC collection efficiency. Such sample sizes allowed testing of appropriate compounds over several hours.

New septa were cut from silicone rubber for each capillary tube, pierced with a needle, and the appropriate end of the capillary tube inserted securely into the glass framework of the olfactometer as shown in Figure 1a.

Stimuli concentrations for sensory testing were set so that the stimuli did not seem extraordinarily weak to panel members with low sensory acuity. Approximately 15 min was allowed for equilibration, then the concentration in each line was measured by GLC and adjusted until the desired digital integrator count ± 5% was obtained. To monitor any line of the olfactometer, the heated stainless steel tube was attached at the "tee" for 30 sec of line purging, the sampling valve injected 0.5 mL onto the column. Results were printed out automatically on a Teletype.

2. Sample presentation Sensory evaluation of odor differences between chiral isomers was conducted using a non-directional duo-trio difference test. Two lines of the olfactometer delivered one isomer while the third line delivered the other. The reference odor sample was presented in the middle nose cone labeled "R" with the two dissimilar chiral isomers presented following a predetermined random protocol in nose cones labeled "A" and "B". Each judge was instructed to sniff "R", "A", "R", then "B". The judge then marked which odor sample, "A" or "B", was unlike "R". In addition, each judge rated the subjective intensity of the odors.

Five judgements or sets were taken each session. After the judge marked the score sheet for one set, the operator connected the nose cone lines to a vacuum. The GLC sampling line was connected to one olfactometer line. After 30 sec, the vapor sample was injected, and the nose cone lines reconnected. Any deviations from desired concentration levels were corrected by slight changes in the microvalve settings. After 30 sec the judge was again asked for an evaluation.

3. Panel selection and testing environment. Six to fifteen judges tested each pair of chiral isomers. Judges were personnel within the department chosen without regard to special olfactory acuity. They were given an instruction session on test procedures and preliminary sessions with the compounds to be tested.

All testing was conducted between 1 to 5 p.m. The olfactometer was next to a fume hood in an unoccupied laboratory. The lighted hood contained the nose cones and was partitioned from the olfactometer and the rest of the dimly lighted room by a screen.

4. Effect of stimulus concentration on discrimination. To establish if levels at which discriminaton between samples might occur due to differences in concentration alone, a series of experiments was conducted using R-(-)-carvone. In the first experiment the carvone was presented at the same concentraton in all three nose cones. In four subsequent experiments the level of one of the non-reference samples was lowered by increments. Twelve experienced judges were asked to pick out the sample unlike the reference in odor type or intensity.

E. Preparation of Chiral Isomers.

1. Synthesis of samples. R-(-)-carvone (specific rotation -60.0 deg) and S-(+)-carvone (specific rotation +59.7 deg) obtained from Fritzsche, Dodge & Olcott, Inc. were purified by preparative GLC and used for testing. Other compounds tested were synthesized from these carvone samples.

a. Carvotanacetone and carvomenthone. The enantiomeric carvones were each reduced by the general catalytic reduction procedure of Brown and Brown (1962). A 250 mL Erlenmeyer suction flask with a balloon wired onto the side arm, containing 70 mL ethanol, 3.6 mL 10% aqueous $PtCl_4$ and 3.6 g powdered activated charcoal was magnetically stirred. A 22 mL portion of stabilized 1 M $NaBH_4$ solution (3.2 g $NaBH_4$ and 0.6 g NaOH in 80 mL ethanol was added and after 10 min, 29 mL concentrated HCl was introduced, followed by 2 mL carvone. With a serum stopper wired over the mouth of the flask, another 22 mL of stabilized 1 M $NaBH_4$ solution was slowly injected through the septum. After 30 min, the flask was gently heated in a water bath over a magnetic stirrer-heater for 15 min. Pressure was released from the balloon by inserting an open syringe through the septum. The solution was filtered, then diluted with 200 mL water, and extracted with ether. The ether extract was washed with 5% $NaHCO_3$ solution, dried over anhydrous $MgSO_4$, filtered and concentrated under nitrogen. This concentrated ether extract was then injected onto the GLC for isolation and purification for sensory analyses.

b. Cis- and trans-dihydrocarvones. The enantiomeric carvones were each reduced by a variation of the Wallach reduction. A 500 mL round bottom flask containing 100 mL distilled water, 50 g powdered Zn, 25 g KOH, 20 mL carvone, and 150 mL ethanol was refluxed under a condensor for 4 h in a water bath over a magnetic stirrer-heater, then filtered. A 200 mL portion of water was added and the solution extracted with ether. The ether extract was successively washed with 5% NH_4Cl, 5% $NaHCO_3$, distilled water, then dried over anhydrous $MgSO_4$, filtered, concentrated under nitrogen and injected onto the GLC for separation and purification for sensory analyses.

2. Purification and identification of samples. Many GLC conditions and columns were tested for preparative scale purification of the compounds. For general preparative separation of most compounds tested, an 8 m Carbowax 20M packed column gave best results. When necessary, other columns were used which gave better resolution of specific compounds. GLC conditions are given in Table 2.

TABLE 2. Conditions for Preparative (top) and Analytical (bottom) GLC

Instrument:	Varian 202 GLC with TCD
Filament current:	125 mA
Carrier gas:	Nitrogen, 50 mL/min
Injector temp:	175 deg C
Detector temp:	230 deg C
Column temp:	100 deg C to elutution temp at 4 or 6 deg C per min
Column type:	6 mm i.d. packed
Instrument:	Varian 1200 GLC with FID
Carrier gas:	Nitrogen, 6 mL/min
Injector temp:	175 deg C
Detector temp:	230 deg C
Column temp:	75 deg to 200 deg C at 2 or 4 deg/min
Column type:	Open tubular, 175 m x 0.75 mm i.d.

Infrared (IR) spectra were taken between sodium chloride plates on a Perkin Elmer Model 237 spectrophotometer equipped with a 5x beam condensor for microsamples. All spectra were run on neat samples purified and collected from a GLC. Mass spectra were obtained using a Varian M-66 spectrometer in a batch mode; samples were introduced directly into the inlet system using the capillary tubes in which they were collected from the GLC. Optical rotation, at the sodium D-line, of the carvones obtained from Fritzsche, D & O was established on a Carl Zeiss Polarimeter with circular scale reading to 0.01 deg using a 100 mm tube with 0.96 mL capacity.

In preparation for sensory evaluation, each compound was collected in 1 to 2 ul amounts from appropriate GLC columns. Compounds were trapped from the exhaust ports of the GLC as originally described by Jennings et al. (1964). Between 6 and 9 ul of these collections were then pooled and injected onto the SF 96(50) packed column shortly before needed for sensory evaluation. Final collections were made in the glass wool filled tubes described previously.

No difficulty was encountered in purification or identification of the carvones or carvotanacetones. An attempt was made to include the carvomenthones in this study, but was abandoned since none of the columns available achieved a satisfactory resolution of sufficient amounts of the cis and trans isomers necessary for sensory evaluation. A 150 m x 0.75 mm SF96(50) capillary, resolved the peaks, but with shouldering. A 5% TCEP packed column resolved the two dihydrocarvones, but when one isomer was collected, subsequent reinjection on the Carbowax 20M capillary column showed approximately 1-2% of the other isomer to be present (see Figure 5). An enol-keto tautomerism, facilitated by the hot surfaces of the GLC, injector port and detector, probably accounted for equilibration of the two isomers.

The stereoisomeric configurations of the dihydrocarvones were established through reduction to their alcohols which were identified by comparing their IR spectra with those published by Mitzner et al. (1968). The two isomeric dihydrocarvones were separated on the Carbowax 20M packed column and reduced on a microscale. Each dihydrocarvone was eluted from the capillary collection tube into a 5 mL centrifuge tube with 0.5 mL ether. A small amount of $LiAlH_4$ was added until the reaction appeared complete, then any excess $LiAlH_4$ was destroyed by adding 4 drops of 10% hydrochloric acid. The solution was extracted twice with 0.5 mL ether. This ether extract was concentrated under nitrogen and injected into the GLC.

The reduction product of the first d-dihydrocarvone collected from Carbowax 20M yielded two major peaks on the Carbowax 20M column which were identifid by their IR spectra as neodihydrocarveol and dihydrocarveol. In a similar manner, the second d-dihydrocarvone collected from Carbowax 20M was reduced to isodihydrocarveol and neoisodihydrocarveol. The two isomers of dihydrocarvone were therefore assigned configurations shown in Scheme 1.

These assignments were reaffirmed by oxidizing isodihydrocarveol and neoisodihydrocarveol and neodihydrocarveol and dihydrocarveol to their respective ketone precursors. Four d-dihydrocarveols obtained from Glidden-Durkee were separated on Carbowax 20M and identified by their IR spectra. The pairs were oxidized on a microscale according to the method of Brown and Garg (1961). The alcohols were eluted from the capillary collection tubes into a 5 mL centrifuge tube using about 0.5 mL ether. A few drops of chromic acid solution (5.0 g $Na_2Cr_2O_7$, 3.8 mL 96% H_2SO_4 diluted to 25 mL) was added while the tube was shaken gently. After 2 h the upper ether layer was removed with a pipet and the lower aqueous layer was extracted twice with 0.5 mL ether. The combined ether solutions were washed with 0.5 mL saturated $NaHCO_3$ then concentrated under nitrogen and injected onto the Carbowax 20M column. IR spectra of the peaks collected were compared to the two dihydrocarvones and structures assigned as shown in Scheme 2.

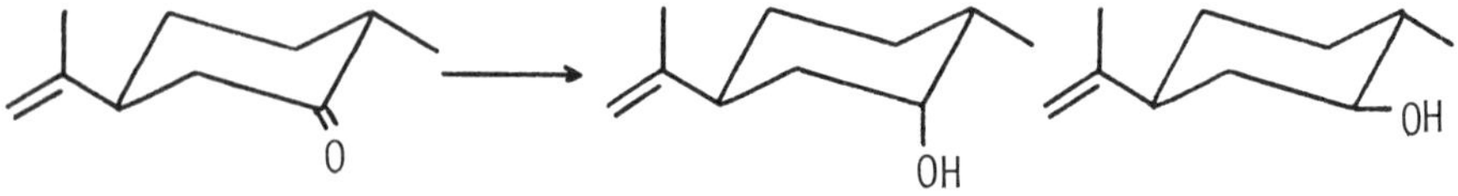

trans-dihydrocarvone (peak 1) | neodihydrocarveol | dihydrocarveol

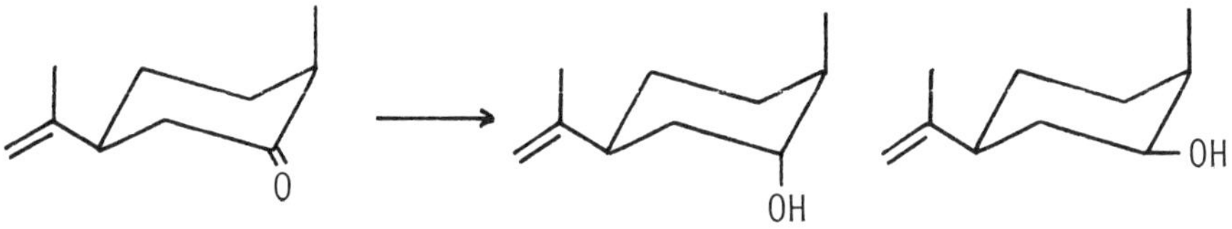

cis-dihydrocarvone (peak 2) | isodihydrocarveol | neoisodihydrocarveol

SCHEME 1. Reduction of the dihydrocarvones.

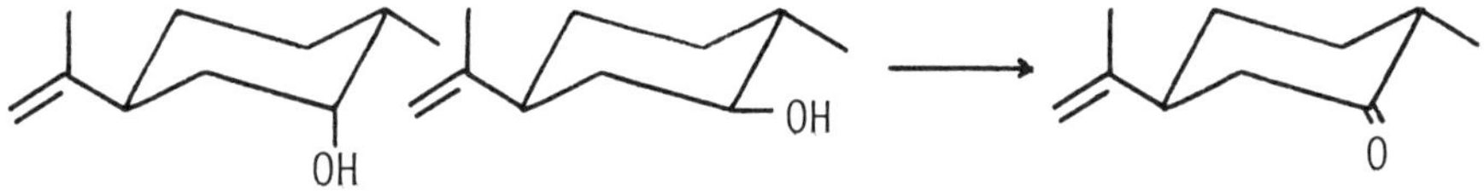

neodihydrocarveol | dihydrocarveol | *trans*-dihydrocarvone (IR identical to peak 1)

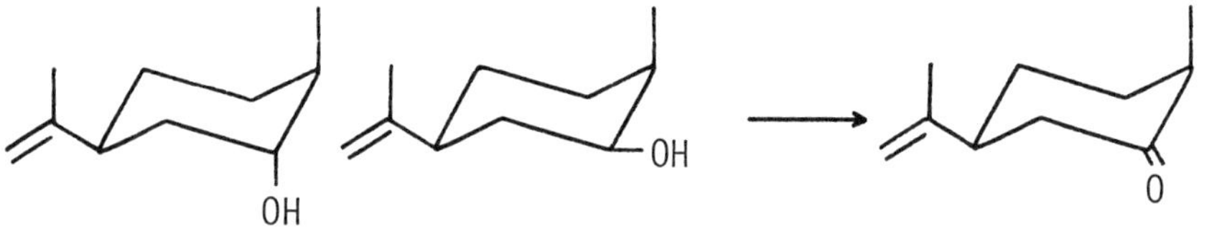

isodihydrocarveol | neoisodihydrocarveol | *cis*-dihydrocarvone (IR identical to peak 2)

SCHEME 2. Oxidation of the dihydrocarveols.

RESULTS AND DISCUSSION

A. The Olfactometer

Sample mass vs. digital response corrected for solvent responses was plotted as shown in Fig. 2. Since 0.5 mL sample loops were used to measure vapor concentrations; multiplying by two gave the concentration in ng/mL of nitrogen. Digital response was a linear function of sample mass over the range used for sensory analysis, as shown in Figure 2. This, and similar calibration curves, were used to calculate the concentrations of stimuli in tests for odor

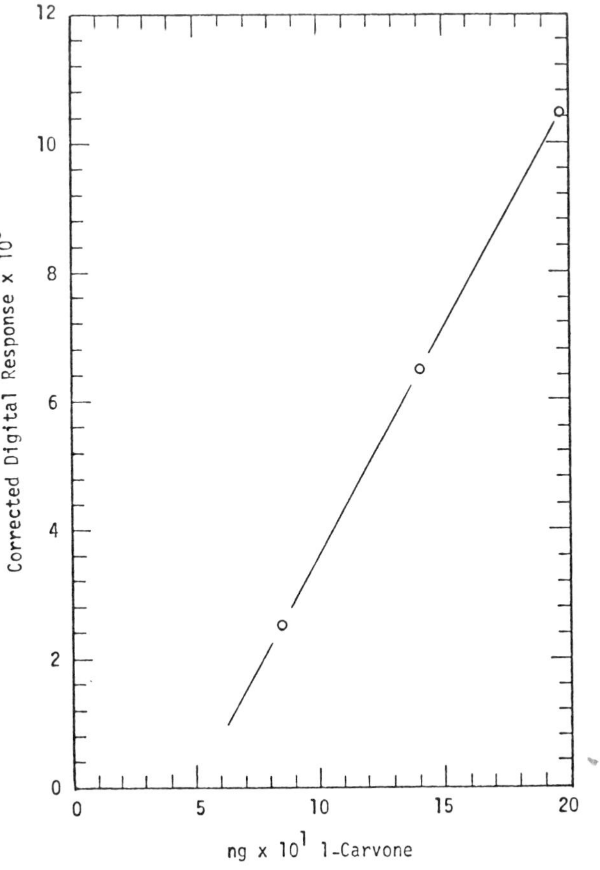

Fig. 2. Detector response to carvone

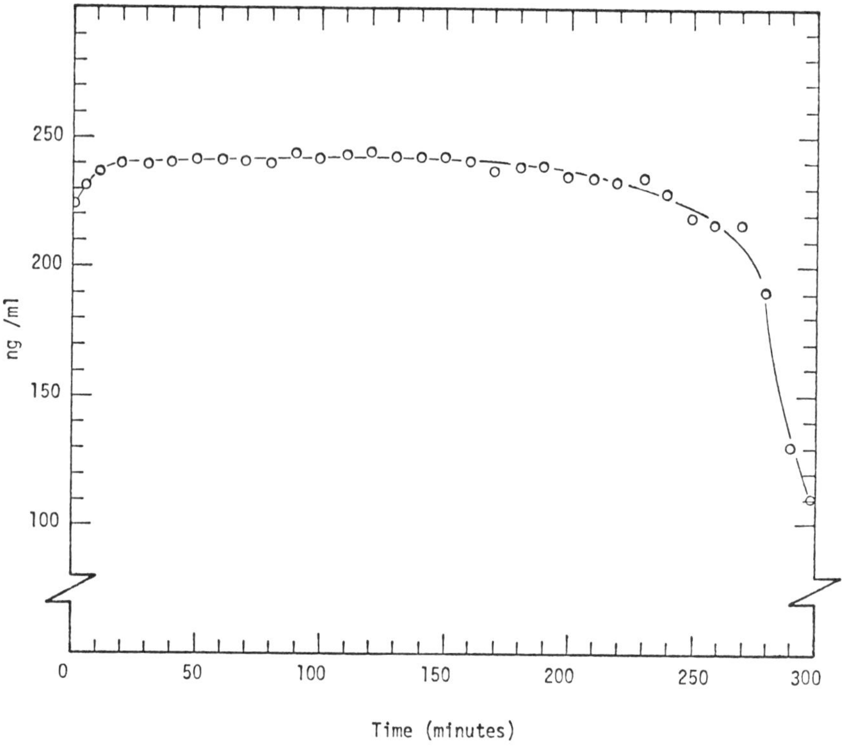

Fig. 3. Monitored carvone concentration vs. time.

differences between chiral isomers in tests for detection of concentration differences, and in determinations of stimuli duration.

The preliminary study of stimulus duration demonstrated that a few mg of sample in the olfactometer was capable of delivering concentrations of stimuli above threshold for several hours.

As shown in Figure 3, stimulus concentration remained nearly constant for approximately 3.5 h, after which it declined rapidly. Once flow rates had been established and the system equilibrated, very little valve adjustment was necessary to maintain a stable concentration. A slight decline in concentration could easily be detected by the GLC monitoring and corrected by appropriate valve adjustments when necessary. Until the rapid fall-off in concentration was reached, the stimulus concentration in the three lines was easily maintained with less than five percent variation.

Precautionary measures for excluding contaminants in the system were investigated and utilized during this study. In routine tests for possible contaminants, judges were asked if they could detect the odor of oil or other impurities in the purified nitrogen carrier gas; they were unable to do so. Sample presentation was randomized by switching delivery lines at the "tees" as outlined above. Judges were unable to detect any residual odors in these lines after vacuum had been applied to the lines between judgements. Odorant stimulus was allowed to purge through the sampling line to the GLC for 30 sec before shunting the 0.5 mL sample onto the column for quantitative analysis. Subsequently, when pure nitrogen was shunted through this line at the same flow rate for 30 sec and the gas analyzed by GLC, no peak response was detectable. It was concluded that 30 sec purging of the sampling system was sufficient.

It is possible that the concentration of odorant in the sampling loop differed slightly from that delivered to the nose cones since the sampling valve line had a greater flow resistance than the nose cone lines. For example, a flow meter placed between the "tee" and the sampling valve or between the "tee" and a nose cone showed flow rates for the same valve settings of 94 and 127 mL/min, respectively. Although the back-pressure created by the constricted flow altered the flow rate, it is uncertain if it appreciably altered the concentration of odorant in the carrier gas. Since the flow through the capillary and main carrier lines was decreased proportionately, any error in measurement of concentration actually delivered to the nose cones would be small. The most important function of this olfactometer arrangement, reproducible delivery of equal concentrations of stimuli in several lines, was not hindered by any such error since any differences were the same in all lines.

B. Sensory Analyses

A sensory experiment on concentration differences was undertaken to determine the levels needed for judges to perceive a significant difference. Figure 4 shows that a significant difference (with $p < 0.05$) was not detected until a 30% or greater change in concentration had occurred, at the stimulus levels in this study. Since the levels of odorant stimuli delivered to the panelists were controlled with less than 5% variation, it was concluded that sensory differences between chiral isomers was not due to perception of concentration differences.

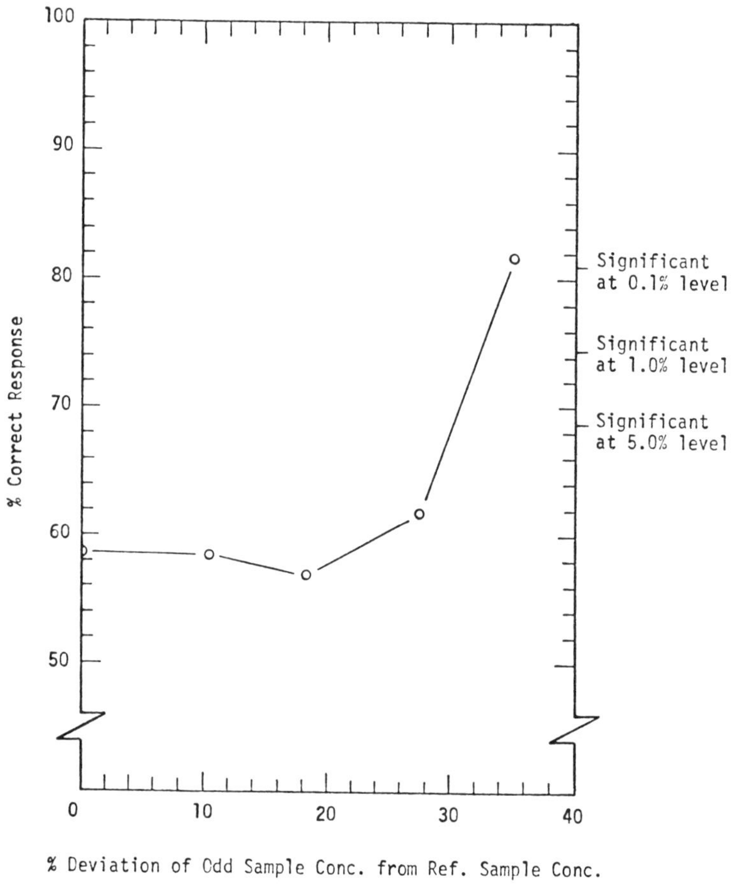

Fig. 4. Detection of concentration differences with carvone.

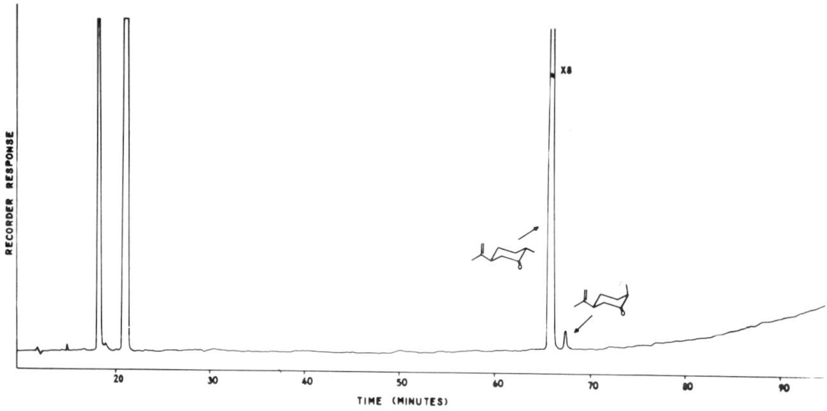

Fig. 5. Chromatogram of trans-dihydrocavone on a 150 m x 0.75 mm Carbowax 20M capillary

The question of sample purity is always of concern. The enantiomers of carvone and carvotanacetone used for sensory analysis were gas chromatographically pure to the limits of detection. The dihydrocarvones, after GLC purification and being re-chromatographed, showed a trace amount of the other stereoisomer, as shown in Figure 5 for trans-dihydrocarvone. Rickborn (1962) studied the cis-trans equilibrium between the stereoisomers of carvomenthone. He found the reaction to be facilitated by either acid or base. It is likely that the similar dihydrocarbones are subject to an enol-keto tautomerism, which in this case might be facilitated by the hot metal surfaces of the GLC analytical tract. It is likely, however, that these minute amounts of alternate stereoisomer were present at levels well below threshold. As discussed below, the cis- and trans-dihydrocarvones have markedly different odors; judges with high sensory acuity did not detect any indication of contributions from stereoisomers other than than those being tested. Even with a subthreshold effect due to the alternate stereoisomers, judges still established differences between the enantiomeric forms of each.

The sensory panels for this study were drawn from laboratory personnel with judges selected without regard to acuity toward the chemicals being tested. Approximately 60 evaluations were done for each compound. Data for the sensory evaluations for each of the four chiral isomer pairs tested were analysed statistically by computing the percent correct response and chi-square value adjusted for one degree of freedom (Amerine et al., 1965). These statistical data along with an outline of synthesis and purification steps are presented graphically for each compound in Figures 6, 7, 8, and 9. Significance was determined at appropriate levels of probability using the chi-square values in tables given by Amerine et al. (1965). For each pair of chiral isomers tested, differences in odor between enantiomers were noted and found to be highly significant ($p < 0.001$). The highly significant differences between each chiral pair show conclusively that sensory differences were perceived by the "average population" represented by the untrained judges. Highly trained judges selected for high sensory acuity would probably record an even greater percentage of correct responses.

Certain trends in odor quality were apparent. The R-(-)-carvone had an odor quality generally described as spearmint; the S-(+)-carvone had an odor quality generally described as caraway. The C4-(R)-enantiomers of carvotanacetone and *trans*-dihydrocarvone, had the spearmint note with the C4-(S)-enantiomers of both carvotanacetone and *trans*-dihydrocarvone described as caraway. The *cis*-dihydrocarvones were variously described, but not as the familiar spearmint and caraway character; however, judges did establish significant differences between the enantiomers.

The possibilities of trace contaminants affecting odor perception with these compounds becomes vanishingly small when one enantiomer is chemically converted to the other with concomitant change of odor. Following the scheme given by Friedman and Miller (1971; see fig 10), interconverted R- and S-carvones were purified by the GLC methods described above, and evaluated in an identical sensory procedure. Results of sensory evaluation (fig 10) were similar to the carvone standards evaluated earlier in the study. If a trace contaminant had cochromatographed with the carvones through all the work-up and purification chromatography, it is not likely

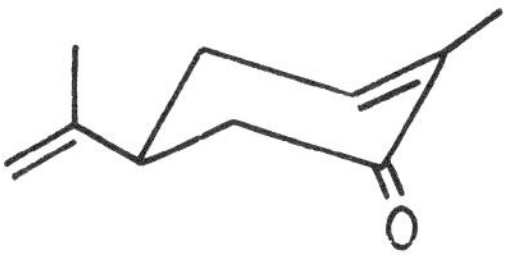

4S-(+)-carvone

&

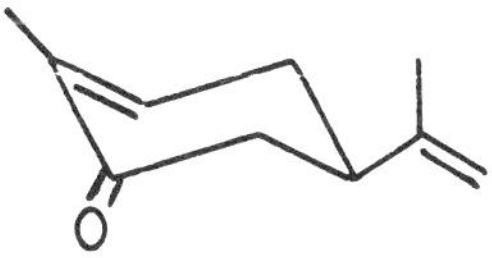

4R-(-)-carvone

Collected from 8 m x 6 mm i.d.
Carbowax 20M column

↓

Collected from 8 m x 6 mm i.d.
SF96(50) column into
glass wool collection tube

↓

Olfactometer

Sensory Analysis:

Duo-Trio Difference Test

Adjusted $\chi^2 = 24.0$ ($P < 0.001$)
% Correct Response = 89.3

FIGURE 6. Preparation and sensory analysis of carvones.

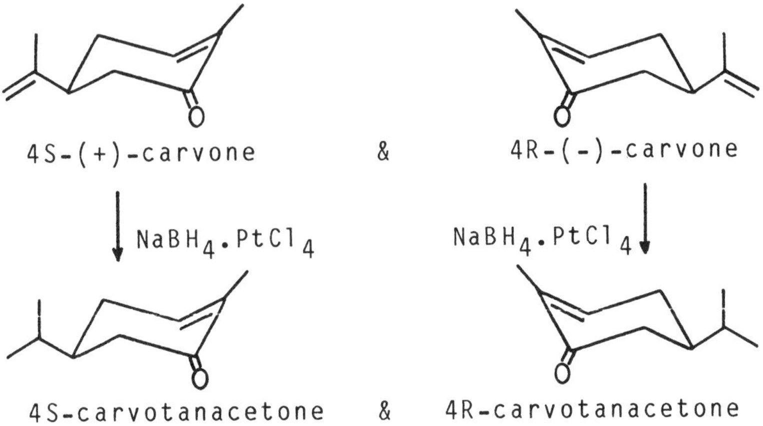

Collected from 8 m x 6 mm i.d.
Carbowax 20M column

↓

Collected from 8m x 6 mm i.d.
SF96(50) column into glass
wool collection tube

↓

Olfactometer

Sensory analysis:

Duo trio Difference Test

Adjusted $X^2 = 13.6$ ($P < 0.001$)
% Correct Resonse = 81.5

FIGURE 7. Preparation and sensory analysis of carvotanacetones.

4S-(+)-carvone & 4R-(-)-carvone

↓ Zn/KOH   Zn/KOH ↓

1S,4S-dihydrocarvone (trans) & 1R,4R-dihydrocarvone (trans)

Collected from 8 m x 6 mm i.d.
5% TCEP column

↓

Collected from 8m x 6 mm i.d.
SF96(50) column into glass
wool collection tube

↓

Olfactometer

Sensory analysis:

Duo trio Difference Test

Adjusted $X^2 = 20.0$ ($P < 0.001$)
% Correct Resonse = 91.7

FIGURE 8. Preparation and sensory analysis of trans-dihydrocarvones.

4S-(+)-carvone & 4R-(-)-carvone

↓ Zn/KOH Zn/KOH ↓

1S,4S-dihydrocarvone (<u>cis</u>) & 1R,4R-dihydrocarvone (<u>cis</u>)

Collected from 8 m x 6 mm i.d.
5% TCEP column

↓

Collected from 8m x 6 mm i.d.
SF96(50) column into glass
wool collection tube

↓

Olfactometer

Sensory analysis:

Duo trio Difference Test

Adjusted X^2 = 11.9 (P < 0.001)
% Correct Resonse = 86.0

FIGURE 9. Preparation and sensory analysis of <u>cis</u>-dihydrocarvones.

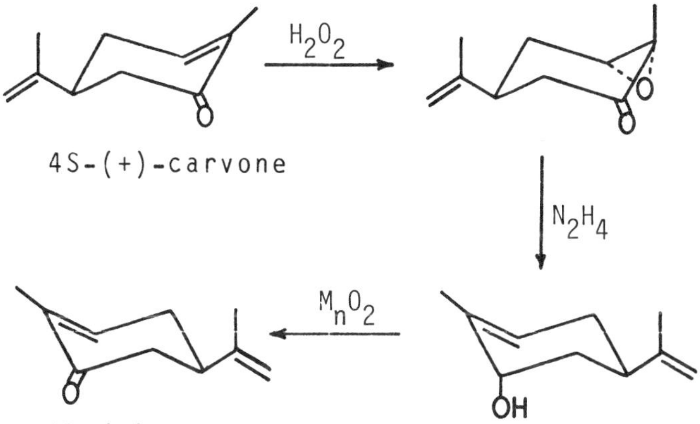

4S-(+)-carvone

4R-(-)-carvone

Collected from 8 m x 6 mm i.d.
Carbowax 20M column

↓

Collected from 8m x 6 mm i.d.
SF96(50) column into glass
wool collection tube

↓

Olfactometer

Sensory analysis:

Duo trio Difference Test

Adjusted $X^2 = 20.8$ ($P < 0.001$)
% Correct Resonse = 89.2

FIGURE 10. Interconversion of carvone chiral isomers.

that it would have changed its odor and thus the difference in odors must have been due to sensory responses to the different chiral isomers.

The data collected in this study support the claim that the human olfactory mechanism is able to discriminate between some chiral isomers. Much further work in the area of olfaction is needed to elucidate the detection mechanism. Studies on the subthreshold contribution of volatiles to odor quality and intensity, the effect of moisture levels in the carrier gas, and the effect of different carrier gases on odor would be of value.

SUMMARY

An olfactometer capable of utilizing small amounts of sample collected from a GLC was developed. An auxiliary unit consisting of an FID GLC digital integrator and Teletype allowed rapid, direct quantitation of odorant in the carrier gas stream of the olfactometer.

The system described was used to present to a panel of judges the erantiomers of a series of stereoisomers (carvone, carvotanacetone, cis-dihydrocarvone, and trans-dihydrocarvone) which had been purified by GLC methods. A duo-trio difference test was used to determine differences in response to the enantiomers. The panel found highly significant ($p < 0.001$) differences in odor between all pairs of enantiomers.

REFERENCES

Amerine, M. A., Pangborn, R. M., and Roessler, E. B. (1965). "Principles of Sensory Evaluation of Food." Academic Press, New York and London.

Bozza, G., Calearo, C., and Teatini, G. P. (1960). Acta Oto-laringologica 52, 189.

Brown, H. C. and Brown, C. A. (1962). J. Am. Chem. Soc. 84, 1493.

Brown, H. C. and Garg, C. P. (1961). J. Am. Chem. Soc. 84, 2952.

Eyferth, K. and Krueger, K. (1969). Institut fur Psychologie der Technischen Hochschule Darmstadt, Germany.
Friedman, L. and Miller, J. G. (1971). Science. 172, 1046.
Jennings, W. G., Creveling, R. K., and Heinz, D.E. (1964). J. Food Sci. 29, 730.
Johnston, J. W. (1967). In "Olfaction and Taste II," (T. Hayashi, ed), Pergamon Press, Oxford and New York.
Mitzner, B. M., Mancini, V. J., Lemberg, S., and Theimer, E. T. (1968). Applied Spectroscopy 22, 34.
Ough, C. S. and Stone, H. (1961). J. Food Sci. 26, 452.
Russell, G. F. and Hills, J. D. (1969). AGFD-6. Am. Chem. Soc. Fall National Meeting.
Russell, G. F. and Hills, J. D. (1971). Science 172, 1043.
Stone, H., Pangborn, R. M. and Ough, C. S. (1965). Advances in Food Research 14, 1.
Stuiver, M. (1960). Acta Oto-laryngologica 51, 135.

COMPUTER ASSISTED QUANTITATION OF CARROT VOLATILES

Gerald F. Russell

Department of Food Science & Technology
University of California
Davis, California

INTRODUCTION

Many studies to characterize the flavor and aroma of fruits and vegetables have used gas-liquid chromatography (GLC) as an analytical device to evaluate the volatile constituents. The sensitivity of GLC methods depends on the detection method and the amount of volatiles present in the foodstuff. To study low concentrations of volatiles in some vegetables, the reproducible and quantitative isolation for GLC analysis is difficult. This chapter reports a method for quantitative GLC analyses which provided a convenient assay; however, the volume of data thereby produced was large. A computer constructed chromographic profiling method was used to assist in the evaluation of these quantitative data.

LITERATURE REVIEW

Carrot (Daucus carota L. var. sativa) is a fleshy storage root in the parsley family (Umbelliferae) of dicotyledonous plants. It is a cool season crop and the variety most widely used for the fresh market is type Imperator, approximately 25 cm in length and 4 cm in diameter.

Carrots are widely recognized for their high vitamin A content, but are rarely purchased if their texture, color, and flavor is poor. A study of the correlation of carrot

CHEMISTRY OF FOODS AND BEVERAGES:
RECENT DEVELOPMENTS

ISBN 0-12-169080-6

textural and sensory properties was reported by Howard and Heinz (1967). Major pigments in carrots include alpha- and beta-carotene with minor amounts of phytoene, lycopene, and xanthophyll; these pigments are probably associated with the presence of beta-ionone in oxidized and fresh carrots (Ayers et al., 1964; Buttery et al., 1979).

Buttery et al. (1961) reported the use of GLC analysis with dual flame ionization detectors (FID) for evaluating the aroma of dehydrated potatoes, carrots, and pears. This method involved the direct injection of the vapor above hot aqueous samples. The same sample preparation method was used by Flath et al. (1969); however, they employed a precolumn trapping system for preliminary concentration of the volatiles. Even though most of the higher boiling fractions could not be detected, this technique proved useful for variety comparisons, for juice blending and for evaluation of aroma loss and change during processing operations.

Interest in the aroma of fresh carrot root led to a preliminary investigation of carrot seed oil from Daucus carota L. (Seifert et al., 1968). The compounds were isolated from different varieties of carrot seeds and the composition of carrot seed oil varies from variety to variety. Differences in reported composition reflect not only varietal differences but also changes occurring in the oil owing to the different methods of isolation of these compounds (Seifert et al., 1968).

The volatile constituents of carrot root oil were reported and reviewed by Buttery et al. (1968). Qualitative and quantitative chemical differences among the volatiles of the carrot roots were not reported. The sample preparation required large amounts of carrots (7 kg) to obtain a suitable amount of sample (0.28 g) and was relatively complicated. Subsequently, further studies have increased the knowledge of volatile constituents of carrots (Heatherbell et al., 1971; Murray and Whitfield, 1975; Cronin and Stanton, 1976; Seifert and Buttery, 1978; Buttery et al., 1979; Simon et al., 1980).

Carrot flavor is dependent on the volatile materials, with quite important contributions from the high boiling fraction. For example, carotol is believed to be a major aroma compound of carrot root. However, steam distillation methods for carrot volatiles, such as described by Buttery et al. (1968) as well as the head space analyses described by other workers,a are not efficient in recovering these higher molecular weight fractions. Buttery et al. (1968) reported that a large portion of the oxygenated fraction of the volatiles was not being detected by their GLC analytical methods and conditions. Simon et al. (1980) reported the use of porous polymer traps to sample carrot volatiles.

There has been interest in some constituents as toxic compounds. Crosby and Aharonson (1967) reported the isolation and identification of a toxic volatile "carotatoxin" as trans-1,10-heptadecadiene-5,7-diyn-3-ol present at levels of about 2 mg/kg of carrot root. Bently and Thaller (1967) corrected this proposed structure, which they identified as falcarinol, first isolated from another Umbelliferae by Bohlmann et al. (1966). Buttery et al, (1968) confirmed the existence of falcarinol in carrot root oil.

EXPERIMENTAL PROCEDURES

A. Instrumentation.

Two GLC instruments were used in this study. The first, a "preparative" GLC was a Varian model 90-P3 with a thermal conductivity detector (TCD). The second GLC, an "analytical" instrument, was a Varian model 1200 with FID and was connected in series with the preparative GLC. The preparative GLC was used to separate the volatile compounds from the solvent and nonvolatile materials; subsequently, the analytical GLC was used to perform a high resolution analysis of the volatile constituents.

The preparative GLC contained a 6 m x 5 mm i.d. glass column packed with 80/100 mesh uncoated glass beads (Corning GLC 110). A glass injector insert, packed with Chromosorb G, 70/80 mesh and treated with dimethyldichlorosylane (DMCS) was used to collect the extractable nonvolatile components. The solid support in the injector insert was frequently changed to prevent nonvolatile components from degrading to artifact volatiles and contaminating the analysis.

The glass bead column was connected through a heated (225 deg C) Carle microvolume valve to the analytical GLC containing a 150 m x 0.75 mm, i.d., stainless steel open tubular capillary column coated with SF96(50). For consistency, SF96(50) was chosen as the liquid phase, following the work of Buttery et al. (1968) on carrot volatiles. Operating parameters for these units are shown in Table 1.

Table 1. Operating parameters for the series connected chromatographs

Preparative GLC:	Varian model 90-P3 containing a glass column (6 m x 5.0 mm id) packed with uncoated glass beads (80-100 mesh)
Helium carrier gas	25 mL/min
Injector temperature	150 deg C
Detector temperature	250 deg C
Column temperature	30 deg C then, 230 deg C
Connecting lines	225 deg C
Microvolumne valve	225 deg C
Analytical GLC:	Varian model 1200 with a stainless steel capillary column 150 m x 0.75 mm id, coated with SF96(50)
Nitrogen carrier gas	6 mL/min
Injector temperature	200 deg C
FID:	
Hydrogen	30 mL/min
Nitrogen make-up	24 mL/min
Temperature	250 deg C
Column temperature	
Initial	70 deg C
Final (and held)	165 deg C
Rate	0.5 deg C/min

B. Sample Preparation.

Carrot seed oil was purchased from Fritzche, Dodge & Olcott, Inc., New York. Carrots (*Daucus carota* L. var. *sativa* type Imperator) were purchased from a retail market on the day each experiment was performed. Both bunched carrots with their tops on and topped prepackaged carrots were studied. Prior to the extraction of the carrot volatiles, the carrots were stored in a refrigerator at 4 deg C.

C. Extraction Procedures.

The carrots used for each experiment were cleaned with distilled water to remove any soil and small root hairs were also removed. After wiping off the excess water, the carrot sample was grated and mixed thoroughly. Approximately 10 to 20 g of this grated carrot were weighed precisely prior to solvent extraction.

The solvent used for the extraction (pentane/carbon disulfide, 20:1, v/v) contained nonane, undecane, tridecane, and heptadecane as internal standards in the amounts of 3, 3, 3, and 6 mg/L of solvent, respectively. All solvents used were chromatoquality reagents. The internal hydrocarbon standards were from PolyScience Corporation.weighed sample of grated carrot, the solvent was titrated quantitatively from a buret in amounts proportional to the weight of the carrot sample (generally 2.00 mL/1.00 g). The mixture was then homogenized in a Virtis blender at the highest speed for 30 sec, and the extracted sample was then centrifuged at 1,240 x g for one min. The pentane phase was decanted and stored in a freezer prior to analysis in vials tightly sealed with teflon-lined screw caps.

D. Injection Procedure.

With the valve directing effluent from the first chromatograph through the loop to atmosphere (top, Figure 1), approximately 300 ul of pentane/carbon disulfide carrot root extract was injected onto the preparative GLC precolumn at 30 $\pm$ 2 deg C. The precolumn was held at this temperature until the solvent had eluted. The trapping loop was then heated for 30 sec using a heat gun to insure that most of the solvent was removed from the loop. The trapping loop was then cooled for 30 min in a Dewar flask containing a dry ice-ethanol slurry. The precolumn was rapidly heated to 230 deg C while the volatile compounds condensed and were trapped in the loop.

After 30 min of trapping, the collected volatiles were injected onto the capillary column by switching the Carle microvolume valve to allow the nitrogen gas to flow through the loop into the capillary column (bottom, Figure 1) while the trapping loop was heated with a heat gun for at least one min. The analytical column oven was then programmed linearly from 70 to 165 deg C at 0.5 deg C/min and held at 165 deg C.

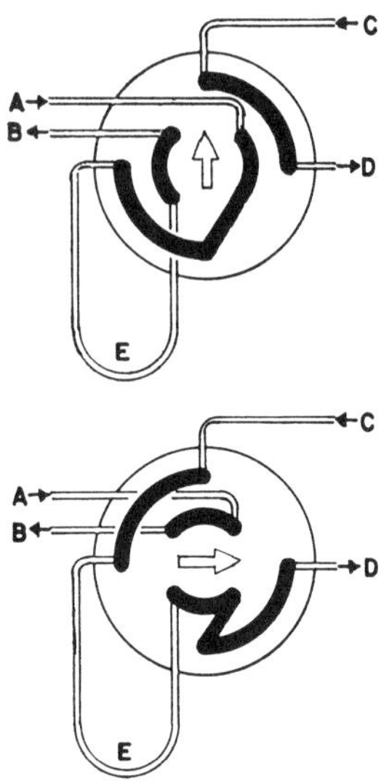

Figure 1. Flow diagrams for connecting valve between series connected chromatographs: A ,line from packed-column chromatograph; B, vent to atmosphere; C, carrier gas supply for capillary chromatograph; D, inlet line of capillary chromatograph; E, trapping loop.

E. Data Treatment.

Gas chromatographic peaks were digitized by an Infotronics CRS-104 integrator. The integrator was interfaced to a Teletype which gave retention time and digitized peak area on tape. These results were analyzed by computer to report the relative retention times, area percentages, average area percentages, and standard deviations and discard very small peaks and random noise.

Computer programs were written in FORTRAN which were specific to the analysis of data from this study. Replicate analyses were processed to give mean retention times, relative retention volumes, mean integrated areas, and standard deviations for all integrated peaks. Since the number of integrated peaks was large, analyses of these data can be tedious and these tabular data will not be given in this paper. However, the use of consecutive plots of chromatographic profiles gives an easily intepreted visual representation of these data. It is the intent of this paper to stress the use of this profiling method.

In addition, a program was written to enter the stored data from integrated analyses and plot a simulated chromatogram under computer control of a Cal-Comp plotter model 750. Simulated Gaussian curves were specifically selected for the particular analysis to match peak width and retention times of the original chromatograms, but they could also be altered to aid in interpreting some analyses. Individual "chromatograms" were then plotted succesively behind each other to generate a three-dimensional, topological display from which differences among the plotted chromatograms were readily discernible. Internal standards in the plotted chromatograms were flagged for computer analyses and all capillary chromatograms were normalized with respect to n-tridecane. A detailed description of our program was given by Glaser et al. (1977).

F. Identification of Volatile Components.

Compounds identified by their retention times, mass spectra and in some cases, infrared spectra were obtained from appropriate fractions and compared with literature values and from key studies by researches outlined in the literature review above. The identication of volatiles followed and confirmed the earlier studies reported above, and comparison with our chromatographs was facilited by the consistent use of the nonpolar GLC liquid phase SF96(50) in most studies. The identification of specific volatiles will not be discussed further here, as this paper concentrates on the methodologies for quantitation and display by computer profiling.

Each constituent collected for IR analysis was purified by preparative GLC using a 1.5 m x 5 mm i.d. glass column, packed with 4% OV-17 on acid washed (AW), DMCS treated Chromosorb G, in a Varian model 202 instrument equipped with a TCD. The IR spectra obtained from the purified compounds were compared with spectra from known standards or reported spectra (eg., Mitzner et al., 1965) for authentication prior to obtaining retention data with the unknowns from carrots. Relative retention times (RRT) relative to the n-tridecane internal standard are given for some compounds to reference position on the chromatograms and also to compare with previous studies reviewed above which used the same liquid phase for GLC analyses.

RESULTS AND DISCUSSION

A. Development of the GLC Systems.

This system employed two series connected gas chromatographs. Figure 2 shows the normalized analyses from quadruplicate runs (as displayed by the plotter) of the carrot seed oil samples injected directly into the 150 m x 0.75 mm id capillary column coated with SF96(50). The standards, nonane, undecane, tridecane, and heptadecane diluted in pentane, were mixed with the carrot seed oil prior to injection. A 0.2 ul aliquot of this mixture was injected directly into the analytical instrument column for each analysis.

Figure 3 shows normalized plotter chromatograms of quadruplicate analyses of carrot seed oil samples injected onto the preparative instrument and subsequently analyzed on the capillary column. These carrot seed oil samples were diluted into 250 ul carbon disulfide/pentane prior to injection onto the preparative GLC. The standards, nonane,.undecane, tridecane, and heptadecane were added, but not in equal portions in the four determinations and were therefore not expected to appear in proportional concentrations. The mean of the relative retention times of volatiles from both sets of carrot seed oil analyses were essentially identical to three significant digits. Comparison of figures 2 and 3 shows that the direct injection, or precolumn solvent stripping method gave essentially the same chromatographic profiles.

Preliminary experiments showed each of a series of liquid phases (SE30, SF96(50), Triton X-305, and Versamid 900) to be unsuitable as liquid-phases for the precolumn. The SE30, SF96(50), and the Triton X-305 liquid phases on various supports produced enough breakdown products to give extraneous peaks in preliminary carrot seed oil analyses. Versamid 900 did not allow high boiling components of the carrot seed oil to elute from the preparative instrument. Any stainless steel columns tried, regardless of which liquid phases or solid support materials, caused rearrangement and subsequent loss of some compounds in the carrot seed oil. Results obtained using glass columns and a glass injector showed greatly improved stability of the carrot seed oil volatiles.

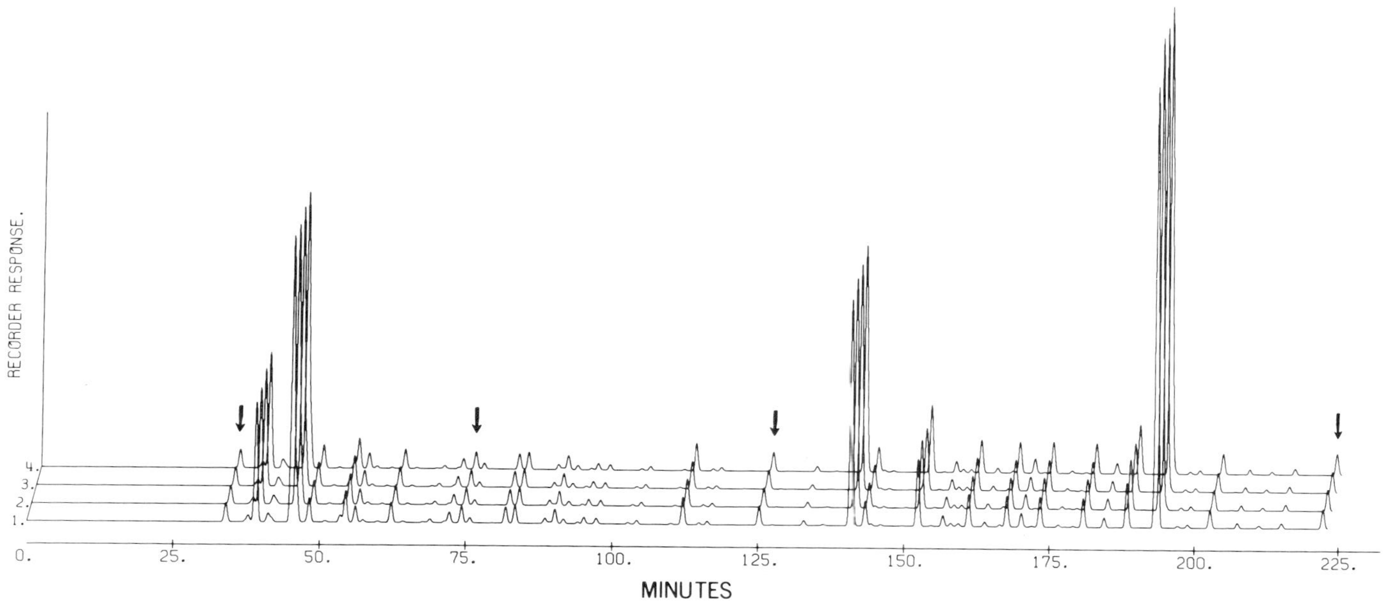

FIGURE 2. Normalized chromatograms of direct injections of carrot seed oil. Arrows point to the internal standards.

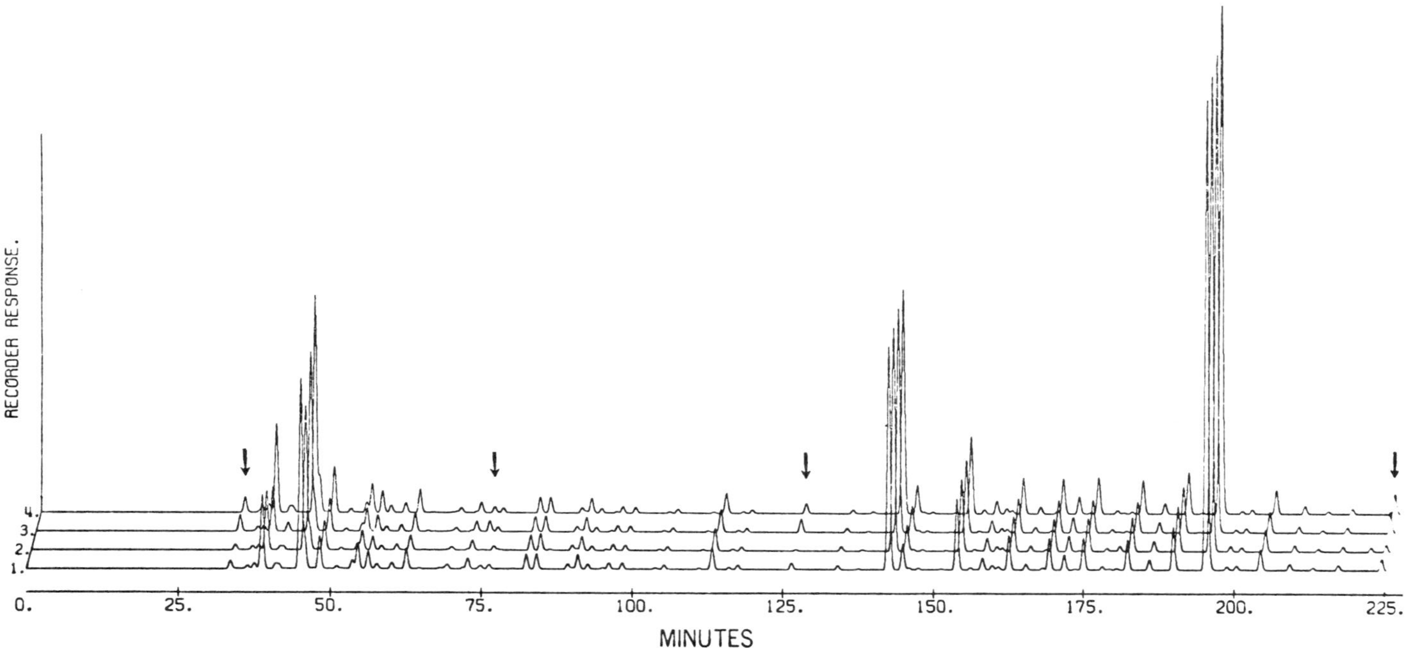

FIGURE 3. Normalized chromatograms of precolumn solvent stripped carrot seed oil dissolved in pentane/carbon disulfide. Arrows point to the internal standards.

Diatomaceous solid supports for the columns also caused loss of some carrot seed volatiles, mostly from the high boiling fraction. Therefore, non-coated glass beads (80-100 mesh, Corning GLC 100 support) were used in the glass precolumn. Because there was no liquid phase to hold up either the lower boiling carrot volatiles or the internal standards, a 6 m column was required to separate the solvent from the compounds of interest in this study. However, the glass injector insert was packed with uncoated 70-80 mesh Chromosorb G to provide a larger surface area on which to collect nonvolatile extracted lipids. This insert was changed frequently to prevent leaching of nonvolatile lipids onto the glass beads in the 6 m column and to prevent their thermal degradation products from producing artifact peaks on subsequent analyses.

Several solvent systems were studied to determine the most useful solvent for this analytical method. Pentane:carbon disulfide (20:1, v/v) was finally selected. Solvents more polar than pentane, e.g., ether, were rejected because they extracted enough water to interfere with the analysis. However, 5% carbon disulfide in pentane was found necessary to obtain a more quantitative extraction of the more polar volatiles found in the carrot root.

Nonane, undecane, tridecane, and heptadecane were used as internal standard markers in each analysis. Nonane was chosen because it elutes from the capillary column before most of the significant carrot volatiles reported in the literature. Heptadecane eluted from the capillary column after all other volatiles in both the carrot seed oil and carrot root oil. These markers were chosen as internal standards for observing the reproducibility of trapping and reinjecting samples during the analysis through both instruments. If the ratio of nonane to other standards was low, there was probable loss of other low boiling volatiles; if the hepatadecane was low, there was a probable loss of high boiling volatiles. Either of these conditions would indicate faulty extraction or injection procedures. The undecane and tridecane standards were chosen as intermediate boiling internal standards since they did not co-chromatograph with major carrot volatiles. All standards were dissolved in the solvent at levels of 3 to 6 ppm. At this level, they produced peaks of appropriate heights for comparison with the carrot root volatiles. For the analysis of the carrot root volatiles, the solvent containing the standards was added by exact volume according to the exact weight of the carrot root sample. Thus, the internal standards enable one to calculate the relative amounts of any one volatile component from any carrot sample and to compare these amounts with those observed from other samples. Simon et al. (1980) reported a porous polymer trapping method for carrot samples, but noted that lower boiling compounds may have been lost and higher boiling compounds may not have been fully represented.

B. Volatile Profiles from Carrots.

Figure 4 shows the normalized plotter chromatograms obtained from four extracts of the volatiles from a single carrot. Each of these chromatograms was from an injection of 250 ul of pentane/carbon disulfide extract of 125 mg of carrot. The standard deviation of the four hydrocarbon standards varied between 1.1 to 2.2 percent of their average area. These chromatograms show very good reproducibility of the extraction and analytical procedures. However, the percentage of standard deviations of some of the individual carrot volatiles varied between 3.9 to 17.7%, reflecting wide variation in volatile contents of individual carrots.

To determine if this technique of analyzing carrot volatiles was sensitive enough to observe differences or similarities in the carrot volatiles between carrots or within a single carrot, three experiments were performed. The first experiment was a GLC analysis of the carrot root volatiles of six different carrots. The second experiment was an analysis of volatiles obtained from 15 acropetal sections cut along the length of a single carrot. The third experiment was a volatile analysis of carrot root samples prepared by dividing the carrot root into the xylem region, the cambium region, and the phloem region. In each experiment, the carrot sample was grated and then extracted with the pentane/carbon disulfide solvent system containing the four internal standards.

Figure 5 shows the normalized plotter chromatograms of duplicate extractions of each of the six whole carrots. The figure shows good reproducibilty for each duplicate extraction of the same carrot. Variations of the mean peak areas between different carrots, however, showed considerably larger differences than differences in replicate runs on the same carrot.

To study variation in carrot volatiles along the length of a single carrot, a carrot was cut into 15 acropetal sections from the bottom of the root to the top, including

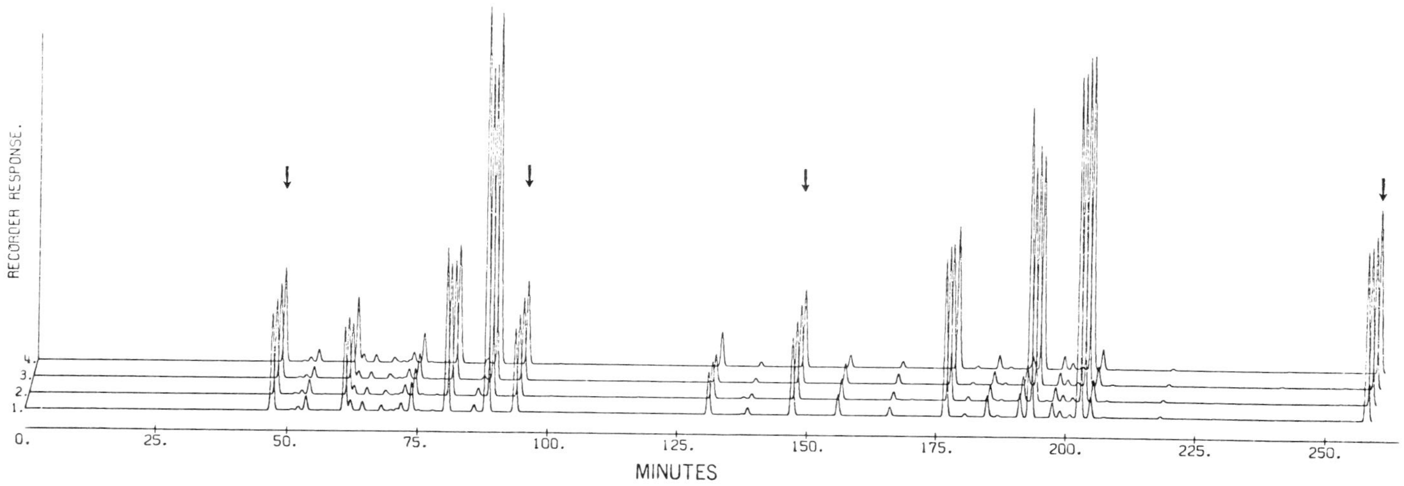

FIGURE 4. Quadruplicate extractions of a single carrot analyzed through dual GLC system.

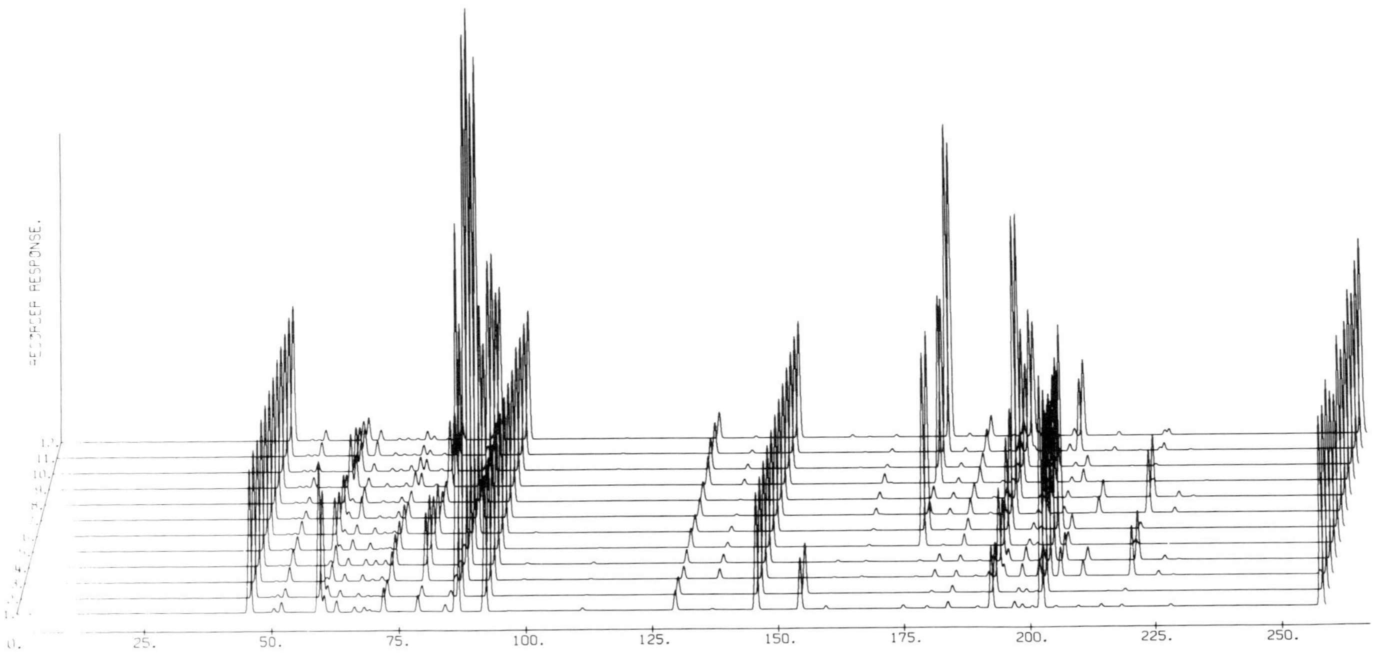

FIGURE 5. Normalized chromatograms of duplicate extractions from six carrots.

approximately 5 cm of green stems in section 15. The carrot was cut in sections of approximately 10 g each. These sections were then grated and exactly 2.00 mL of solvent with standards were added by titration for every 1.00 g of carrot for extraction.

Figure 6 shows the normalized plotter chromatograms of these 15 analyses. It should be noted that the top chromatogram (15) in Figure 6 represents the top section of the carrot which was mostly green stem. The two major peaks in this section, sabinene (RRT 0.41) and myrcene (RRT 0.44) were reported by Buttery et al. (1968) to be the principle volatile components of carrot tops. These two constituents show a remarkable increase in both the carrot top (15) and in the top section (14) of the carrot root. When comparing the differences among carrot volatiles through the acropetal sections (up to the top sections, 14 and 15), these results show only minor variations along the length of the carrot. Most components are either fairly consistent in level or show random changes as does gamma-terpinene (RRT 0.60). There is, however, one peak (RRT 1.31) which had increased four times on one run (chromatogram 9). This sudden change in the concentration of this single constituent may have been due to its being pooled in this region (although this seems surprising) or its localized biological activity producing this higher concentration.

In order to study variations in the carrot volatiles across the diameter of a carrot, two carrots were divided into the xylem region, the cambium region, and the phloem region prior to grating and extraction for analysis. The normalized plotter chromatograms resulting from analysis of these three regions for both carrots are shown in Figures 7a and 7b, respectively. These figures show that major diferences do occur in the concentration of individual volatile constituents through these three regions. The xylem region (top chromatogram (3) of both samples) shows fewer constituents and lower concentration in the higher boiling point fractions than is observed in the other two regions. The higher boiling constituents are at higher concentrations in the cambrium region (middle chromatogram, number 2 in both samples). These results also reflect the gross differences in the volatile constituents observed between different carrots.

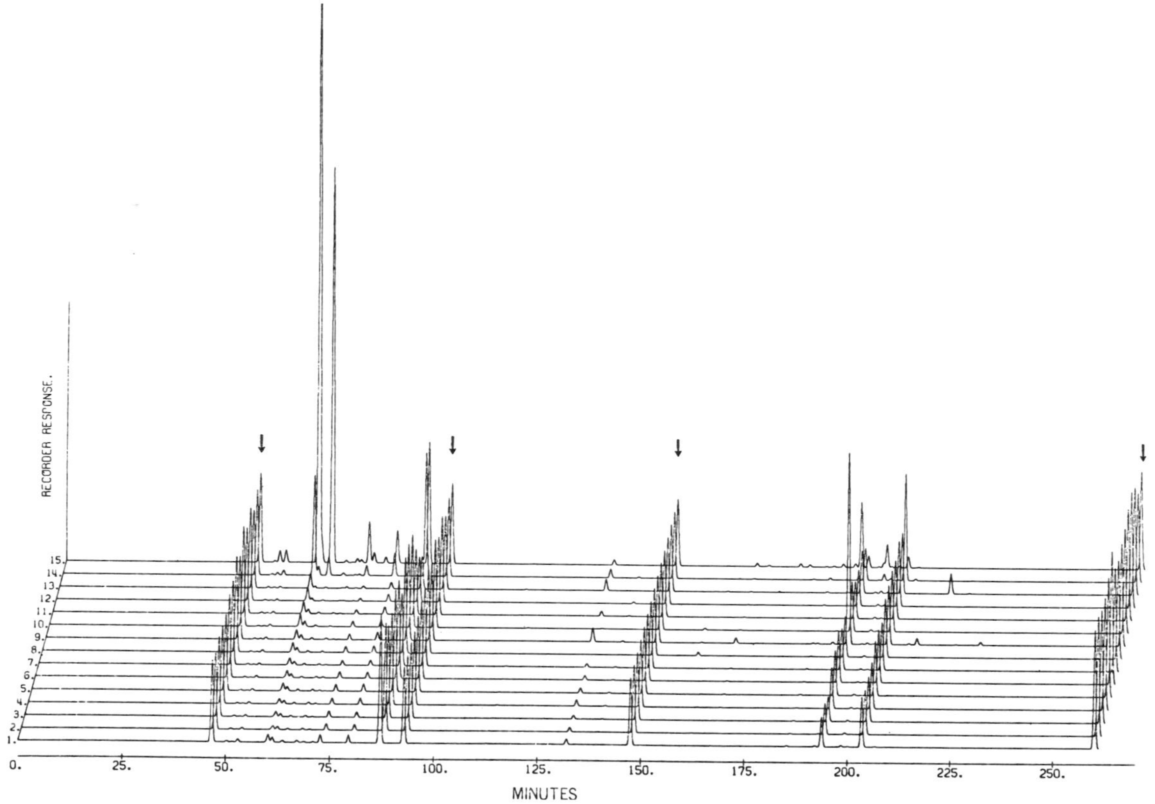

FIGURE 6. Normalized chromatograms of extracts from 15 acropetal carrot sections.

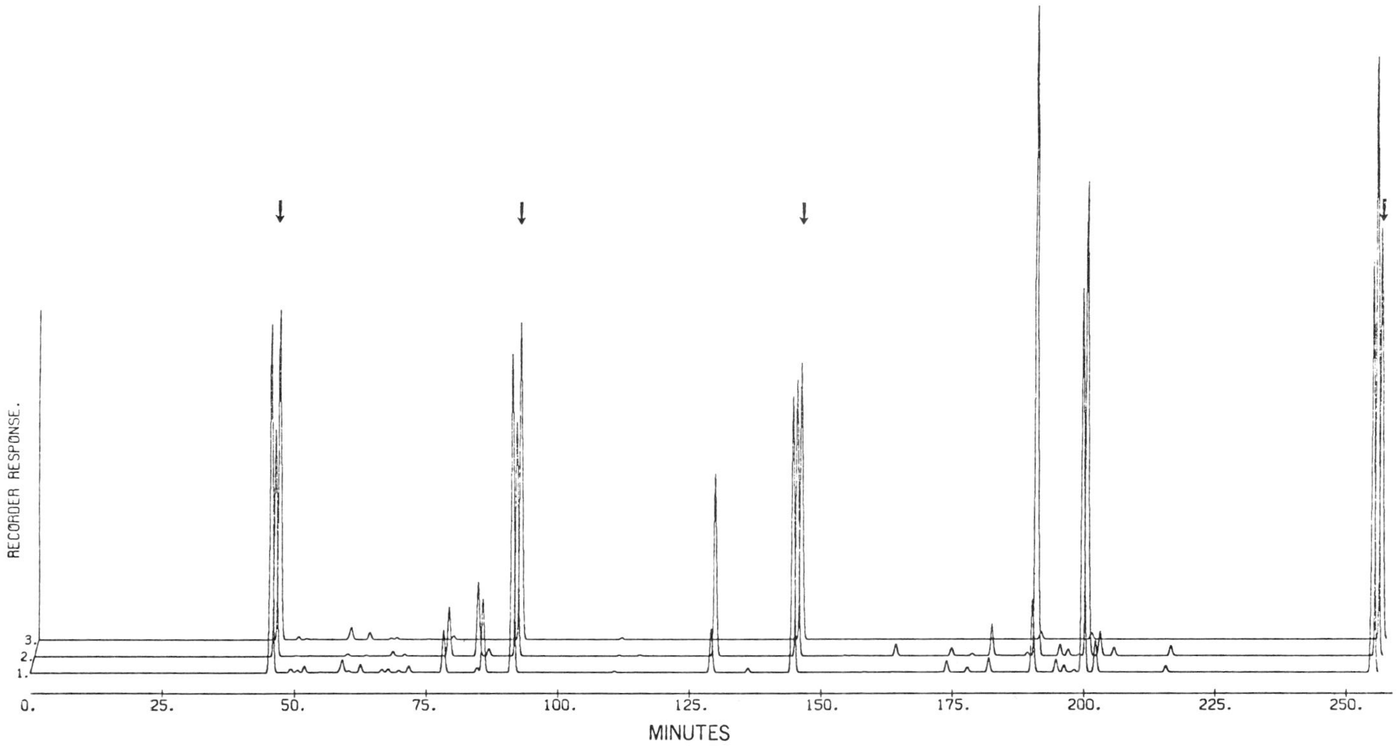

FIGURE 7a. Normalized chromatograms of extracts from the phloem (1), cambium (2) and xylem (3) regions of a single carrot.

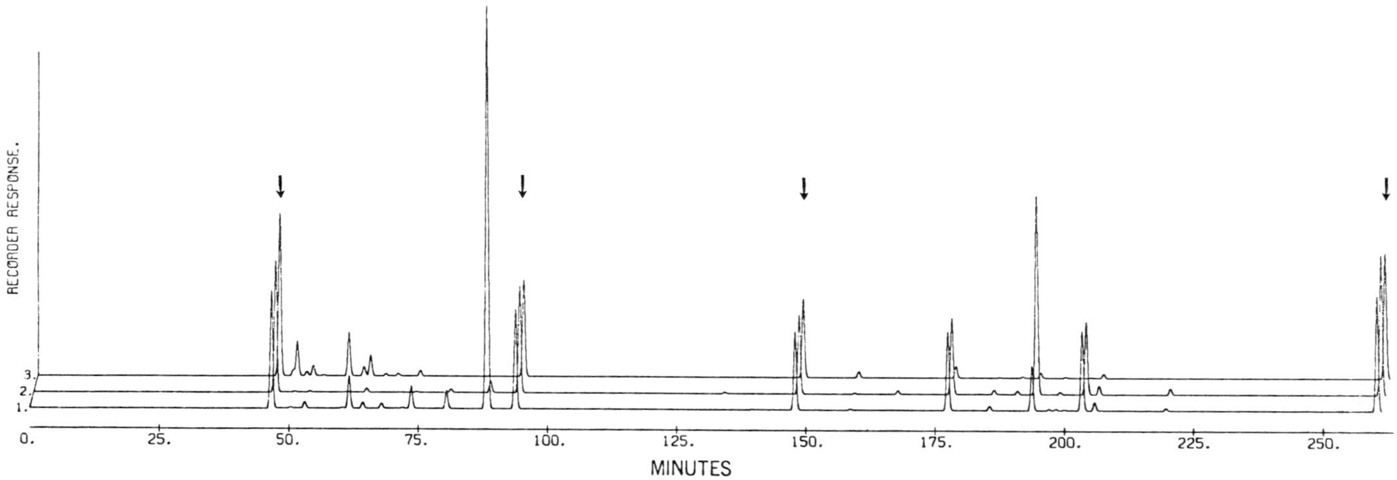

FIGURE 7b. Normalized chromatograms of extracts from the phloem (1), cambium (2), and xylem (3) regions of a second carrot.

SUMMARY

A system for quantitative extraction and subsequent analysis of carrot volatiles by highly efficient GLC methods was established. The isolation of the carrot volatiles involved a direct extraction of the carrot lipids. The separation and analysis of the volatile constituents contained in this lipid extract utilized two GLC instruments connected in series. The first instrument was a preparative chromatograph used to separate the carrot volatiles from both the solvent and the nonvolatile lipids. The second instrument contained a highly efficient capillary column and was used to perform the analysis of the carrot volatiles.

Model systems, using both carrot seed oil and carrot root extract, were employed to determine the reproducibility of the system of analysis and at the same time to determine differences and similarities in the carrot root volatiles between different carrots and within one carrot. The results showed that carrot root volatile analysis can be accomplished utilizing this system. Further, the results showed that there is a gross difference in the composition and the amounts of the carrot root volatiles from carrot to carrot. The results also showed very different concentrations of carrot root volatile components in the xylem, cambium, and phloem regions of a carrot, but differences in carrot volatiles taken along the length of a carrot are much less.

ACKNOWLEDGEMENTS

Early phases in the the development of GLC operating conditions and extraction procedurs were done with able assistance and many long discussions with D. E. Heinz and C. Singhagarn.

REFERENCES

Ayers, J. E., Fishwick, M. J., Land, D. G., and Swain, T. (1964). Nature (London) 202, 81.

Bentley, R. K. and Thaller, V. (1967). Chem. Soc., London, Section D, Chem. Commun. 9, 439.

Bohlmann, F., Niedballa, U., and Rode, K. M. (1966). Chem. Ber. 99(11), 3552.

Buttery, R. G. and Teranishi, R. (1961). Anal. Chem 33, 1439.
Buttery, R. G., Seifert, R. M., Guadagni, D. G., Black D. R., and Ling, L. C. (1968). J. Agric. Food Chem. 16, 1009.
Cronin, D. A. and Stanton, P. (1976). J. Sci. Food Agric. 27, 145.
Crosby, D. G. and Aharonson, N. (1967). Tetrahedron 23, 465.
Flath, R. A., Forrey, R. R., and Teranishi, R. (1969). J. Food Sci. 34, 382.
Glaser, E. R., Silver, B., and Seiffet, J. H. (1977). J. Chromat. Sci. 15, 22.
Heatherbell, D. A., Wrolstad, R. E., and Libbey, L. M. (1971). J. Food Sci. 36, 219.
Howard, P. L. and Heinz, D. E. (1970). J. Texture Studies 1, 185.
Mitzner, B. M., Theimes, E. T., and Fruman, S. K. (1965). Applied Spectroscopy 19, 169.
Murray, K. E. and Whitfield, F. B. (1975). J. Sci. Food Agric. 26, 973.
Seifert, R. M., Buttery, R. G. and Ling, L. (1968). J. Sci. Food Agr. 19, 383.
Seifert, R. M. and Buttery, R. G. (1978). J. Agric. Food Chem. 26, 181.
Simon, P. W., Lindsay, R. C. and Peterson, C. E. (1980). J. Food Sci. 28, 549.

TASTE AND TASTE EVALUATION OF SOY PROTEIN HYDROLYZATES

Jens Adler-Nissen
Hans Sejr Olsen

Enzyme Applications R&D
Novo Industri A/S
Bagsvaerd, Denmark

I. INTRODUCTION

Soy protein products are widely applied in the food industry where they serve as functional ingredients and often also as low-cost sources of dietary protein. The functional properties of soy flours, concentrates, isolates and textured products do not, however, fulfil all existing demands, and further modification is therefore often desirable.

Enzymatic hydrolysis is an established and attractive way of improving the functional properties of food protein ingredients. For example, whipping agents are produced commercially by enzymatic hydrolysis of soy isolate with pepsin (1), and several other enzymatic processes for the improvement of the foaming and/or emulsification properties of soy protein products have been developed (2-5). Enzymatic hydrolysis also results in an increased solubility of the soy globulins at their isoelectric point (pH 4-5), and it appears that most of the hydrolysis processes described in the literature are specifically directed at achieving complete solubility. Such hydrolyzates may find use in soft drinks or in cured whole meat products with a view to an increase of their protein content (6). Important examples will be discussed later in the proper context.

Indiscriminate enzymatic hydrolysis of soy protein and many other food proteins generally results in the formation of bitter peptides, whereby the protein hydrolyzate is rendered more or less unacceptable as a food ingredient. How to avoid a bit-

ISBN 0-12-169080-6

ter taste is unanimously regarded as the most severe problem in the production and use of soy protein hydrolyzates (7), but bitterness is not the only off-flavor which should be considered. In soy flour the protein is associated with beany flavor compounds which are released during hydrolysis (8) having an adverse effect on overall organoleptic quality. There is also a contribution to the taste from possible ions used in pH adjustments during the production of the hydrolyzates. Finally, it should be stressed that it is the organoleptic quality of the food product rather than of the hydrolyzates per se which determines the acceptability of soy protein hydrolyzates as food ingredients, and an examination of the taste of soy protein hydrolyzates must therefore be seen in the light of their end use.

With respect to end use, it is fruitful to distinguish between soy protein hydrolyzates which are applied mainly because of their functional properties, such as the whipping agents described above, and soy protein hydrolyzates which are used for nutritional fortification. The first category, which may be called "functional hydrolyzates", is mostly used in low concentrations in complex food systems, whereby a possible slight off-flavor of the hydrolyzate may be masked. The second category, exemplified by iso-electric soluble soy protein hydrolyzate (ISSPH) (6), is used in relatively high concentrations in order to give any meaningful contribution to an individual's daily protein intake. This results in very critical demands on the taste if ISSPH is to be accepted for protein fortification.

It may be proper to stress that "protein fortification" is understood here in its most simple meaning, namely an increase of the protein content of a food relative to its mass or energy content. However, when discussed in relation to nutritional impact, protein fortification is a most complicated issue, in particular in developing countries (9). In the industrialized countries protein fortification has no relevance for the nutritional status of the general population (10), and it is rather the extension of high cost proteins with low cost proteins which is relevant. This is the rationale behind the incorporation of soy protein in meat products.

However, there are presumably sections of the population in Europe and North America who because of low food intake and/or bad eating habits are subclinically malnourished. This has been shown in at least one instance, namely surgical patients, who because of poor appetite for solid, protein rich food are nearly always in a negative nitrogen balance (11). Their nutritional status is further deteriorating because of a high physiological demand for protein in the period of convalescence, and clinical evidence of the positive effect of extra protein intake has been procured in several cases (11).

On the background of the above, we considered it of value to carry out a systematic taste evaluation of ISSPH. The results of this work can, however, only be properly understood as part of a more general, theoretical discussion of flavor problems of protein hydrolyzates.

II. THEORY

A. Changes in Flavor and Functionality during Hydrolysis

The hydrolytic degradation of protein can be quantitatively described by the term DH (degree of hydrolysis) which is defined as the percentage of peptide bonds cleaved (7). The advantage of describing the reaction by DH and the means of controlling it have been reported in detail (12-14), and it shall only be stated here that DH can be shown to be the principal determinant for the composition and properties of a protein hydrolyzate made from a given protein and enzyme (14).

If a protein is hydrolyzed to different DH-values and the bitterness of the hydrolyzate is evaluated before any further treatments, a qualitative relationship can be established between bitterness and DH as shown in Figure 1. This relationship is discussed in depth elsewhere (4), and only the main conclusion will be given here: The relationship is based on the generally acknowledged observation that bitterness is positively correlated with the hydrophobicity of the amino acid

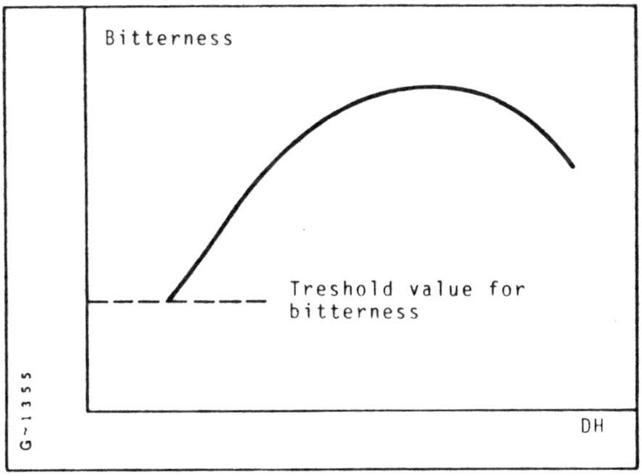

FIGURE 1. Qualitative relationship between bitterness and DH

side chains (15). In the intact globular protein molecule the majority of the hydrophobic side chains are concealed in the interior and therefore cannot interact with the taste buds. When the protein is degraded, the hydrophobic side chains become exposed to the solvent, resulting in the development of bitterness. However, at the beginning of the hydrolysis, the peptides formed are still fairly large, so that they can to some extent mask the side chains by hydrophobic interaction. These peptides are then further degraded, resulting in more and more of the hydrophobic side chains becoming exposed; hence bitterness increases. The bitterness begins to decrease again at very high DH-values, because more and more of the hydrophobic amino acids will become terminal or be released as free amino acids, and these compounds are less bitter compared with peptides in which the same amino acids are in endo-positions.

Figure 1 has deliberately been drawn without units as the quantitative relationship between bitterness and DH may vary enormously from system to system. Even when we consider only one particular protein-enzyme system, the level of bitterness will be strongly influenced by the presence of other compounds, as will be demonstrated later with the addition of malic acid to ISSPH. The question of what DH-value is the limit for bitterness therefore cannot be answered in general.

Concomitant with the development of bitterness changes are observed in the various functional properties of the protein hydrolyzate. Some of these properties, e.g. emulsification capacity and whipping expansion, pass through an optimum early in the hydrolysis reaction, whereas isoelectric solubility increases steadily with increasing DH (4). This is in accordance with the fact that in processes for functional protein hydrolyzates, the hydrolysis reaction is limited in extent, which has the further advantage that bitterness is less pronounced. On the other hand, in processes for production of soluble hydrolyzates such as ISSPH, a more extensive hydrolysis is needed in order to get a satisfactory yield, and this aggrevates of course the bitterness problem.

From the above it is obvious that the production of ISSPH must involve more than mere enzymatic hydrolysis if an acceptable product is to be obtained. All published processes at least include a separation of soluble peptides from still insoluble proteinaceous matter, but most of them are characterized by having additional refinements which are claimed to be crucial for avoiding bitterness. In the following the most important of the process types and their relation to the organoleptic quality will be discussed.

B. Methods for Controlling Off-Flavors in ISSPH

A review of the literature concerning the production of soluble hydrolyzates from soy protein shows a multitude of ingenious methods for controlling the off-flavor of ISSPH, in particular bitterness. The methods can be systematically summarized as shown below:

Starting material:	Defatted soy flour or flakes
Pre-treatment:	- Conversion to concentrate by acid wash - Conversion to isolate by extraction and precipitation
Enzymatic reaction:	- Choice of particular enzyme(s) - Choice of particular hydrolysis conditions
Separation step:	- Selective precipitation - Separation by ultrafiltration
Post-treatment:	- Selective extraction or adsorption - Carbon treatment - Flash evaporation - Plastein reaction - Masking

1. Pre-Treatment. A pre-treatment of the soy flour is advantageous as a means of lowering the characteristic beany off-flavor. Many processes start by making isolate in situ (16-18); however, a conversion to concentrate, as in our process (6), is a more economical way of removing water soluble off-flavor compounds.

2. Enzymatic Reaction. The choice of enzyme is critical for the yield and plays a role for the development of bitterness as well. Different endopeptidases have different specificity; for example Alcalase(R) (subtilisin Carlsberg) is known to cleave preferably at the carboxylic side of hydrophobic amino acids (19), which, according to the model for development of bitterness described in the previous section, should result in less bitterness than if cleavage took place at hydrophilic amino acids. However, a systematic study of the effect of endopeptidase specificity on bitterness has yet to be carried out, as previous works have not separated the effect of enzyme specificity from that of DH (20,21).

If exopeptidases are used together with endopeptidases, some of the bitter peptides may be degraded to non-bitter peptides and free amino acids. This approach has most convincingly been demonstrated for casein hydrolyzates by Clegg (22,23), but work on soybean protein has also been reported (24).

The choice of particular hydrolysis conditions is also important, but only few specifications can be found. Pour-el and Swenson (16) stress the advantage of hydrolyzing under quiescent conditions, and we have specified a narrow range of pH and DH-values (25). DH determines the yield as well as the composition of the ISSPH: The yield increases with increasing DH whereas the average peptide chain length (PCL), which is shown in Figure 2, decreases. PCL is smaller in ISSPH than in the hydrolysis mixture before separation, which is obvious because high molecular weight, insoluble protein is separated from the ISSPH. PCL decreases relatively more slowly in ISSPH than in the complete hydrolyzate, which is a simple mathematical consequence of the increase in yield with increasing DH, and this means that the composition of ISSPH is fairly insensitive to minor fluctuations in DH. The values of PCL at the lower end of the DH range are presumably large enough to allow that hydrophobic interaction between side chains may take

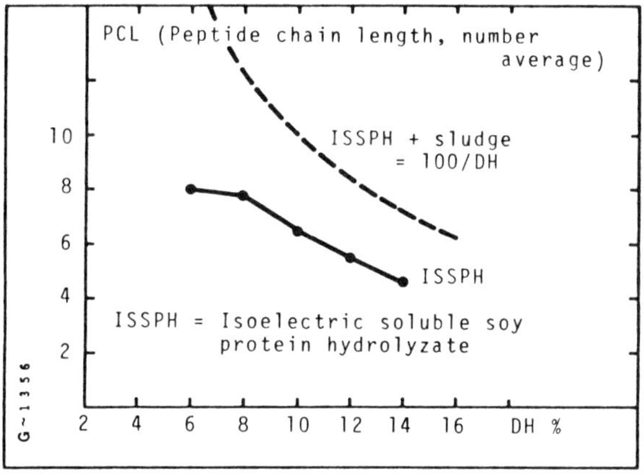

FIGURE 2. Average peptide chain length of ISSPH. The peptide chain length is the number of amino acid residues and is measured as the ratio of total and free α-amino groups in the hydrolyzates (7). Free α-amino groups are assayed by TNBS using a standard curve for soy protein (26). The samples of ISSPH were prepared as described in the present work.

place (4), and it therefore seems reasonable to expect that we are still on the ascending part of the bitterness-DH curve in Figure 1. Thus, we may expect an increase in bitterness with increasing DH. On the other hand, with decreasing PCL the buffer capacity of the ISSPH increases, and this means that an ISSPH with a high DH value contains relatively more acid anion (resulting from lowering of the pH from pH 8 to pH 4.2) than an ISSPH with a low DH value. As we shall see later, the presence of ions influences the sensation of bitterness markedly, and we may therefore expect that the predicted increase in bitterness is not always observed in practice.

3. Separation. The separation step may play an important role in decreasing the bitterness in our process (6), because the hydrolyzate mixture before separation is found to be quite bitter at the relevant high DH values (4). One may hypothesize that the debittering is caused by the precipitation of small, highly hydrophobic peptides, as such peptides would be insoluble at pH 4.2 which is around the isoelectric point. No differences in the amino acid composition of ISSPH and the original raw material have been observed, however (6), and if the hypothesis is correct, it must therefore be a question of minute amounts of highly bitter peptides which are precipitated. Such intensely bitter, highly hydrophobic peptides have in fact been isolated from a soy protein hydrolyzate made with Alcalase(R) and separated at alkaline pH where these peptides are soluble (27).

A completely different way of carrying out the separation is continuous ultrafiltration in a membrane reactor, first and most represented by the work of Iacobucci and co-workers (28). The peptides formed from the described process were very small indeed (29), and it was also found that there was a gradual build-up of insoluble, hydrophobic proteinaceous matter (28). Both phenomena may contribute to the observed low bitterness of the hydrolyzate.

4. Post-Treatment. This step encompasses a considerable number of different methods apart from our own carbon treatment: extraction with azeotropic, secondary butyl alcohol (27), hydrophobic chromatography (27,30) and flash evaporation to remove beany off-flavor (31). In addition to these, the remarkable plastein reaction, which has been extensively studied (32), deserves mentioning. By the plastein reaction a gel is formed from a liquid hydrolyzate at high dry matter concentration using a protease as catalyst. The mechanism is a reversible transpeptidization whereby peptides, which can form a gel by secondary bonding, are synthesized (33).

Finally, the effect of masking must be mentioned. The bitter taste of protein hydrolyzates can be masked by the addition of peptides rich in glutamic acid (34). In our process the presence of anions has a considerable masking effect which differs with the type of ion: If chloride is present, the ISSPH is immediately evaluated as slightly bitter, whereas if the same ISSPH contains malate or citrate, no bitterness can be detected in a simulated soft drink. This will be discussed later in more detail.

III. EXPERIMENTAL

A. Production of ISSPH

ISSPH was produced (both in laboratory and pilot plant scale) from a commercial soy isolate (Purina 500 E) by hydrolysis with Alcalase 0.6 L (35) using the following standard hydrolysis parameters: Substrate conc. = 8.0% N x 6.25; enzyme dosage: 12 Anson Units (AU) per kg substrate protein; pH = 8.0 (pH stat); temperature = 50°C. At a given DH-value the hydrolysis was terminated by the addition of acid (HCl or DL-malic acid) to pH 4.2, whereafter the ISSPH was recovered by centrifugation and filtration as described elsewhere (13). The raw ISSPH was then treated with activated carbon (BGN from Lurgi) for 30 min. at 50°C. Kjeldahl N was measured before and after the carbon treatment. pH for the carbon treatment was pH 5.0, and the carbon dosage was 2% of the protein (Nx6.25) in the raw ISSPH unless otherwise specified. In one case the ISSPH (HCl inactivated) was desalted by hyperfiltration (36). The samples of ISSPH were stored frozen and thawed immediately before the taste panel experiments. The samples are identified throughout this work by their DH-value and the anion, e.g. "DH 10, malate".

B. Preparation of Particular Samples

For some of the experiments described later on, the preparation of the samples necessitated certain modifications of the above procedure.

1. Effect of Enzyme. Three series of ISSPH were produced using the following enzymes and dosages:
a) Alcalase(R) 0.6 L: 6 AU/kg, 12 AU/kg (standard process), 24 AU/kg, 48 AU/kg, and 96 AU/kg.

b) Subtilisin A, a highly concentrated form of Alcalase(R) (37): 6 AU/kg, 12 AU/kg, 24 AU/kg, 48 AU/kg, and 96 AU/kg.

c) Alcalase(R) 0.6 L: 6 AU/kg plus varying amounts of inactivated Alcalase 0.6 L. The inactivation was performed by heat treatment of the enzyme preparation at 95°C for a few minutes, and the enzyme was added to the hydrolysis mixture in concentrations corresponding to 0 AU/kg, 6 AU/kg, 18 AU/kg, 42 AU/kg, and 90 AU/kg.

2. Effect of Anion. The concentration of malate and chloride, respectively, in a given sample of ISSPH after separation of the sludge is roughly the same as before the separation, because the proteinaceous sludge is removed at its isoelectric point where no counter-ions are needed. From the mass of added acid and the Kjeldahl analyses of the ISSPH it is therefore possible roughly to calculate the ratio of malate or chloride to protein in ISSPH. This ratio increases with increasing DH because of the concomitant decrease in peptide chain length (cf. Figure 2). By addition of extra Na^+ and malic acid or HCl to ISSPH with DH 6, it is possible to adjust the salt content to that of ISSPH with higher DH-values.

3. Effect of Cation. ISSPH was produced from acid washed soy flour (6), and KOH was used instead of NaOH for neutralization before and during the hydrolysis. The hydrolysis was stopped at DH 10% with malic acid.

4. ISSPH in Whole Meat Products. ISSPH, "DH 10, chloride", was produced in pilot plant as described above and spray-dried with an inlet temperature of 200°C and an outlet temperature of 90°C. Steam sweeping of the atomizer was used to increase the bulk weight of the powder (38). The powder, which contained 90.1% Nx6.25 on a dry matter basis, was dissolved in water to make a brine with the following composition: 8.2% Nx6.25, 9.5% NaCl, and 1.75% Na-polyphosphate. The brine was neutralized with 4N NaOH to pH 6.7 and incorporated into pork leg at the Danish Meat Research Institute, Roskilde, by the following procedure: 12 kg of mechanically tenderized, sectioned pork leg was massaged with 4.8 kg of the above brine for 20 h at 5°C, whereby the brine was absorbed entirely. The meat pieces were stuffed into sausage casings and cooked. The cold, ham-like product was sliced and served to the panel for evaluation of bitterness.

C. Taste Evaluation

All taste panel evaluations took place in a uniformly lit room equipped with individual booths for the panelists. The ISSPH samples (30 ml) were served in brown, 100 ml glass jars at room temperature. The samples were adjusted to pH 6.5 with NaOH and to 4% Nx6.25 by dilution with water. Chilled water and bland tasting bisquits were available ad libitum.

The samples were evaluated for one or more of the particular flavors: bitterness, soy flavor, salt taste, other off-taste. The methods for evaluation were either ranking (lowest rank = lowest intensity of flavor) or a scale with five scores (1 = no flavor, 3 = medium flavor, 5 = strong flavor, with the option of giving the scores, 2 and 4). In some experiments triangular tests were used for comparison of samples.

In most of the ranking experiments the same series of samples were evaluated twice by the panel; in those cases the data were treated as if obtained from a correspondingly larger number of judges. Such pooling of the data increases the sensitivity of the ranking evaluation and is statistically acceptable as long as it is used in cases like the present, where it is difficult to discriminate between the samples, and particular products cannot be recognized immediately. The ranking data were treated by calculating the coefficient of concordance using the null-hypothesis that the ranking takes place randomly, as described in standard references, e.g. Amerine et al. (39). The data from the scoring tests were treated by analyses of variance, and the statistical tables of Roessler et al. (40) were used for evaluating the results from the triangular tests.

IV. RESULTS AND DISCUSSION

A. Selection and Training of The Taste Panel

The panel was selected and trained according to the principles described by Cross et al. (41): First, the potential candidates (all employees at the research departments of Novo) were informed of the project and invited to participate. Those who gave positive response were then requested to fill in an interview form to establish in particular the candidate's interest and ability in participating. 38 candidates were then able to continue with the screening tests, which briefly consisted of: 1) A test for recognizing 10 common aromas; 2) A test for ability of assigning unknown quinine solutions to known standard solutions; 3) A test for threshold value of quinine; 4) A test for recognizing soy flavor in solutions as well as in meat balls. On the basis of these screening tests 13 panelists were selected and trained. The training program consisted of the following tests: 1) Repeated ranking tests for bitterness and soy flavor of standard quinine solutions in ISSPH (20, 40, 80, and 160 ppm quinine hydrochloride); 2) Scoring tests on the same standard solutions using the scale of 1-5; 3) Ranking tests for soy flavor in soy protein products. In addition to the training tests, ad hoc tests were carried out and discussed with the panelists in order to establish the optimal conditions for serving the samples of ISSPH (pH, sugar addition, etc.). It was found that maximum sensitivity for bitterness was obtained if the ISSPH was neutralized to pH 6.5 and served at room temperature in a concentration of 4% Nx6.25 with no sweetener added. When the ISSPH was served as a simulated soft drink (pH 4.2, 3% Nx6.25 and 9% sucrose), the panel was much less able to note any bitterness.

At the end of the training period the panel was able to rank correctly the standard solutions of quinine in ISSPH, making only insignificant misjudgements (4). The panel was also able to evaluate with reasonable consistency the bitterness level and the level of soy flavor of the standard solutions using the scale. As shown in Figure 3, there is a clear, positive correlation between the bitterness score and the concentration of quinine. The score for soy flavor, on the other

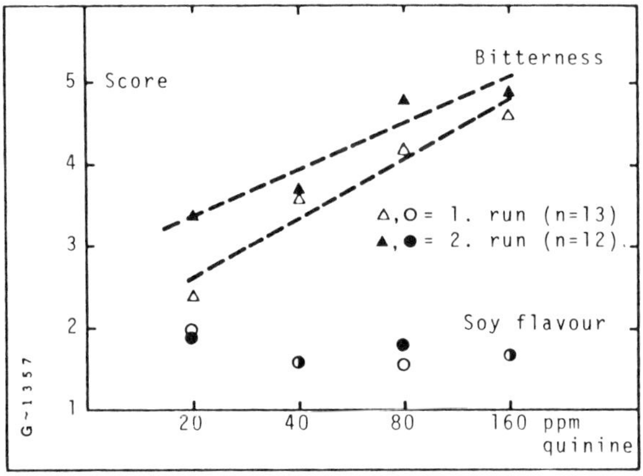

FIGURE 3. Evaluation of standard quinine solutions in ISSPH

hand, is independent of the presence of quinine, which demonstrates the panel's ability to discriminate between soy flavor and general bitterness. The training test shown here was repeated twice with a period of four months in between, and it appears that the score for soy flavor could be reproduced. The scores for bitterness, on the other hand, increased somewhat during this period, which demonstrates the increased sensitivity towards bitterness as a result of regular practising. Thus, at the time of the first test run, the threshold level of quinine for the panel was found to be about 20 ppm, but at the time of the second test this concentration was clearly higher than the threshold value. This has the consequence that scores for bitterness should only be compared for series of experiments which have been carried out within a shorter period of time, e.g. one month. (This requirement is fulfilled for all the scoring tests discussed later on.)

This change in sensitivity, which is observed (not unexpectedly) when using the scoring test, could be circumvented if the samples were ranked together with the standards, to give the quinine equivalent value, QEV (4). Thereby a highly reproducible, quantitative measurement of bitterness is obtained. However, the method is considerably more time consuming and straining for the panel, and it should therefore only be used in cases where a strict quantification is essential.

B. The Effect of Ions

The presence of ions in ISSPH complicates the evaluation of bitterness because of the expected masking effect of particularly malate. The masking effect of malic acid is clearly demonstrated by the following experiment:

To desalted ISSPH, malic acid was added in concentrations of 5, 10, 20, and 30% of the total dry matter (including the malic acid). The samples were then neutralized and ranked according to bitterness. The result was unequivocally clear: The panel agreed highly on the order of the samples with the 5% sample being most bitter followed by 10%; 20%, and the 30% sample being least bitter, in fact not bitter at all. 30% malate corresponds to a little below 50% of the (Nx6.25) content, and this is about the same concentration as found in ISSPH which has not been desalted.

Next, the effect of adding extra salt to a "DH 6, malate" product was studied. Malic acid was added up to a concentration corresponding to a "DH 10, malate" product, and the samples were then ranked according to salt taste. The same experiment was carried out with HCl, too. The results are shown in Figure 4, and it can be concluded that a significant in-

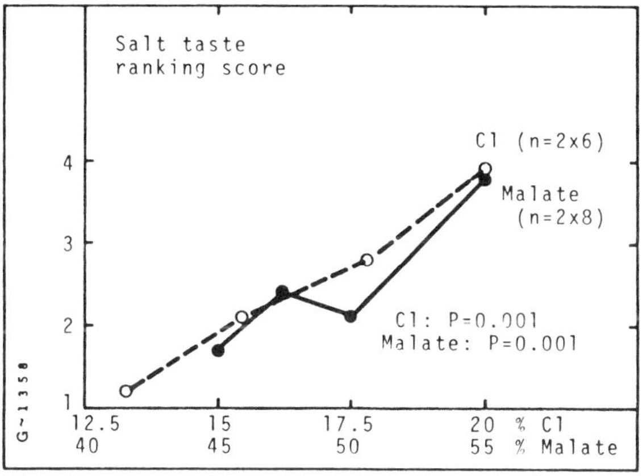

FIGURE 4. Evaluation of added salt to DH 6 ISSPH. The lowest concentration corresponds to the salt already present and the highest concentration to the amount of salt in a DH 10 ISSPH. The concentrations are expressed as % of Nx6.25

crease in salt taste was observed. The same results (data not shown) were obtained also with other acids (citric acid, phosphoric acid) as well as with ISSPH products in the range DH 10 to DH 14 - in all cases a strongly significant (P<0.001) increase in salt taste was observed. This was quite surprising, because the panel was not able to distinguish between "DH 10, malate" and "DH 14, malate" in triangular tests (Table I). As consistent results were obtained when the experiments were repeated, it is beyond doubt that the results are the expression of a genuine phenomenon. The only reasonable explanation is

TABLE I. Triangular tests on ISSPH with different DH-values

Anion	Samples compared		Number of trials n	Correct judgements	Significance level P
Malate	DH 10	DH 14	18	6	n. s.
	DH 10	DH 14	12	4	n. s.
	DH 10	DH 15	16	9	0.05
	DH 10	DH 16	16	8	n. s.
Chloride	DH 10	DH 15	18	10	0.05

G-1429

therefore that not only can malate and other ions mask the bitterness of the hydrolyzate, but the hydrolyzate also has a masking effect on the salt taste, and this masking effect increases with increasing buffer capacity, i.e. DH-value. It is relevant here to mention that masking of salt taste is a well-known phenomenon, for example the masking effect of sucrose is quite pronounced (42).

Having established this mutual masking effect of the salt and the hydrolyzate we had to conclude that it was not possible to study the effect of DH on bitterness isolated from the effect of the ions. We might of course have worked with desalted ISSPH without neutralization, but because the desalination process results in an appreciable loss of nitrogen compounds which are permeating the hyperfiltration membrane (36), it would be difficult to control whether these desalted products are truly representative. The ions are in fact inherent in ISSPH and ISSPH must be studied with those ions present as they occur at the end of the production process.

A final experiment concerned the effect of substituting K^+ for Na^+ in the ISSPH process, because the use of K^+ may have some advantages if the ISSPH is intended for dietary use. Two "DH 10, malate" products were produced from acid washed soy flour using NaOH and KOH, respectively, for the neutralization and pH-stat hydrolysis. A triangular test using a simulated soft drink gave the result that 8 our of 9 could distinguish between the two products ($P<0.02$), and all 8 indicated that the product containing K^+ was bitter. Apparently the bitter taste of K^+ cannot be sufficiently masked by the malate. The use of K^+-containing ISSPH will therefore require some development work before an acceptable product can be obtained.

C. Evaluation of Bitterness

Two series of ISSPH products (one with malate, the other with chloride) covering the range DH 6 to DH 14 were ranked according to bitterness. The results are shown in Table II. No significant difference in the ranking of the malate products could be observed, whereas there was a significant increase in bitterness in the case of the chloride products. This agrees with the observation mentioned previously that the masking effect of malate is much higher than that of chloride. The level of bitterness was not evaluated quantitatively in this series of experiments, but qualitatively it was found, as usual, that the malate products were low in bitterness (barely above the threshold limit), whereas the chloride products exhibited a medium bitter taste. These observations for malate were confirmed by triangular tests (Table I).

TABLE II. Evaluation of Bitterness of ISSPH by Ranking

Acid Anion	Average ranking for samples of ISSPH					Number of judges	Signifi-cance level
	DH 6	DH 8	DH 10	DH 12	DH 14	n	P
Malate	2.8	2.8	2.9	2.7	3.9	2 x 8	n.s.
Chloride	2.8	2.3	3.0	2.9	4.2	2 x 10	0.01

Although a statistically significant increase in bitterness is observed in the case of the chloride products, the difference is slight and presumably of little importance in practice. This is demonstrated by a triangular test on "DH 10, chloride" versus "DH 15, chloride" (Table I). Only a weakly significant difference was observed, and it is obvious that if there were any major differences between the two products, the trained panel should have no problem in reaching an almost unanimous vote.

It has now several times been stated that ISSPH made with chloride exhibits some bitterness. However, this bitterness appears to be easily masked in food products - for example, we were not able to detect any bitterness when a "DH 10, chloride" ISSPH was incorporated into cured pork in a concentration of 2.3% Nx6.25 on the final product weight (see the experimental section for details).

D. Evaluation of Other Off-Flavor

The dosage of Alcalase$^{(R)}$ 0.6 L used in the production of ISSPH may sometimes have to be higher than the standard dosage of 2% (12 AU/kg) in order to reach a high DH-value (e.g. DH 15) in a reasonable time. There is no immediate reason to expect any flavor problems caused by an increase in enzyme dosage, because it has been demonstrated that increasing the enzyme dosage does not change the properties of the hydrolyzate, as long as it is hydrolyzed to the same DH value (14). However, it is obvious that there must be an upper limit to the enzyme dosage at which the flavor of the enzyme product itself becomes noticeable. The flavor of ISSPH (DH 10, malate) produced with the commercial enzyme product Alcalase$^{(R)}$ 0.6 L, which has an activity of 0.6 AU/g, was therefore compared with a similar ISSPH produced with a highly purified preparation of Alcalase$^{(R)}$ enzyme. Because of the high activity (24 AU/g for the particular batch used), only minute amounts are needed to

TABLE III. Evaluation of Possible Off-Flavor in ISSPH originating from Alcalase(R) 0.6 L

Activity dosage AU/kg	Ranking of dosage ↓	Ranking of enzymes → Subtilisin	Alcalase	Inactiv. Alcalase	Number of trials n	Significance level P
6	1.6	1.7	2.1	2.2	3x10	n. s.
12[a)]	2.6	1.7	2.2	2.1	3x10	n. s.
24	2.0	1.8	1.8	2.3	3x11	0.05
48	4.1	1.8	2.6	1.6	3x12	0.001
96	4.8	(distinct taste of Alcalase at 96 AU/kg)				
G~1430	n= 2x10 p= 0.001	a) 12 AU/kg corresponds to standard dosage of 2% Alcalase 0.6L/kg substrate protein				

reach even a large activity dosage.

In a simple ranking according to off-flavor of ISSPH produced with Alcalase(R) in different dosages the results shown in the second column of Table III were obtained. There is obviously a significant difference between the samples, and the figures indicate that the threshold value is somewhere between 24 and 48 AU/kg. This preliminary conclusion was confirmed in the following experiments where the different enzymes were compared at the same activity dosages.

Table III shows the results from ranking of ISSPH produced with subtilisin, ISSPH produced with Alcalase, and ISSPH to which inactive Alcalase was added (cf. the experimental section). At activity dosages of 6 and 12 AU/kg no significant differences could be observed. At 24 AU/kg there was a weakly significant difference which apparently is due to the sample containing inactivated Alcalase. Not until at 48 AU/kg the difference between subtilisin and Alcalase becomes pronounced in accordance with the conclusion above. The high score for ISSPH containing inactivated Alcalase 0.6 L at 24 AU/kg and the subsequent low score at 48 AU/kg indicates that heat inactivation of the Alcalase preparation has changed the flavor of the enzyme preparation in an uncontrollable way, and the conclusion is therefore mainly based on a comparison of the first two rows.

The final series of experiments concerned the important question of the carbon treatment. The treatment with activated carbon improves the flavor of ISSPH considerably, and in initial trials the panel had no difficulty in distinguishing untreated from treated samples. The carbon treatment reduces the soy off-flavor, but it could not be ruled out that it might

TABLE IV. Evaluation by Scoring of The Effect of Carbon Treatment on The Taste of ISSPH

Flavour	pH	Score for pH 2% dosage	Score for carbon dosage 1%	2%	4%	8%	Number of judges n	Significance level P
Soy	4.2	2.7	2.2	3.3	2.6	2.6	11	0.05
	4.5	2.4	2.9	3.1	2.8	3.2	12	n. s.
	5.0	2.1	2.9	2.1	1.6	1.7	10	0.001
	6.5	2.2	-	-	-	-		
Bitterness	4.2	2.6	2.6	2.5	2.2	1.9	11	n. s.
	4.5	3.3	3.0	2.4	2.4	2.1	12	n. s.
	5.0	2.6	2.7	2.8	2.1	2.0	10	0.001
	6.5	2.6	-	-	-	-		
	Avrg.	2.8	2.8	2.6	2.2	2.0		
		n = 9 n. s.[a]						

[a] Both soy flavour and bitterness

G-1428

influence bitterness because of adsorption of hydrophobic amino acids. Under all circumstances it was considered necessary to make a systematic investigation of the effect of the two most significant parameters in the carbon treatment: pH and dosage. The evaluation was made by scoring, because it was felt necessary to establish not only relative differences, but also an absolute level of soy off-flavor and bitterness in the carbon treated ISSPH.

Table IV shows the results of the evaluation of the carbon treatment. Several series of experiments were carried out with either pH or dosage being varied within the same series. The results at 2% dosage agree well with each other which indicates that the scorings are reproducible within the experimental period.

With respect to soy flavor no significant difference is found in the effect of pH at 2% level, although the score at pH 5.0 is consistently lower than that at pH 4.2. Increasing the dosage does not reduce the soy flavor when the treatment is carried out at pH 4.2 or 4.5, but at pH 5 there is a significant decrease in soy flavor with increasing carbon dosage. However, for economical and technical reasons it is not recommendable to use the high dosages in industrial scale, and the result is therefore mainly of theoretical interest: The observed interaction between dosage and pH indicates that there are some charged off-flavor compounds which bind weakly to the carbon. If the compounds were uncharged, it is not likely that this strong effect of pH at high carbon dosages would be observed.

In the case of bitterness, there is no significant effect of pH, and this appears to hold for all dosage levels. A uniform decrease in bitterness is observed with increasing carbon dosage, which justify the calculation of the average scores shown in the last row. The decrease in bitterness can be readily explained by adsorption of hydrophobic peptides. Such a debittering effect of large dosages of activated carbon has been demonstrated for casein hydrolyzates, and it was found that the debittering was accompanied by a selective loss of particularly the hydrophobic amino acids phenylalanine and tryptophan (43). That hydrophobic adsorption does take place is further substantiated by the independency of pH, as no charges are involved in the adsorption. However, as with the soy flavor it is presumably not advantageous to exploit in practice the debittering effect of high carbon dosages.

The scores for both soy flavor and bitterness are all in all reasonably low, and the results indicate that it might even be possible to reduce the recommended carbon dosage to 1% without any deteriorating effect on the organoleptic quality of the ISSPH. It should be taken into consideration that the samples have been evaluated under the most extreme conditions in order to increase the sensation of off-flavor as much as possible. It is, in fact, our general experience that as soon as the samples are not neutralized, but sweetened instead to cover the sour taste, both bitterness and soy flavor disappear altogether in a "DH 10, malate" ISSPH.

IV. CONCLUSION

Both the theoretical considerations and the experimental work lead to the conclusion that there is no particular crucial step in the production process which is responsible for an acceptable organoleptic quality of ISSPH. At a first glance this might be regarded as a drawback because it may seem that the given recommendations should be followed closely. However, the results also show that although each of the steps is important, there is a considerable flexibility in the choice of the processing conditions. For example, DH can be varied much more than in the case of the functional protein hydrolyzates where the development of the improved functionality can be ascribed mainly to DH. There is also a considerable freedom in the choice of the parameters for the carbon treatment, as discussed above.

Another general conclusion which should be drawn from the experimental work is that the fairly bland taste of these ISSPH products is the result of a complex interaction of masking effects from all the components present. This means that the complete composition of the protein enriched beverage should be balanced, not only the components of the ISSPH as such. For example, ISSPH produced with malate has an agreeable acidulous taste in a sucrose solution, but when added to fruit juices the total concentration of organic acid may be too high. Reduction of the malate in ISSPH then seems necessary, for example by desalination and/or partial replacement of the malic acid by HCl.

ACKNOWLEDGMENTS

Our colleagues in the taste panel are warmly thanked for their wholehearted assistance and endurance. These thanks are extended to Mrs. Gudrun Poulsen who skilfully organized the practical details of the taste evaluations.

REFERENCES

1. Gunther, R.C., J. Am. Oil Chem. Soc. 56:345 (1979).
2. Puski, G., Cereal Chem. 52:655 (1975).
3. Zakaria, F., McFeeters, R.F., Lebensm. -Wiss. u. -Technol. 11:42 (1978).
4. Adler-Nissen, J., Sejr Olsen, H., ACS Symp. Ser. 92:125 (1979).
5. Bobalik, J.M., Taranto, M.V., J. Food Technol. 15:637 (1980).
6. Sejr Olsen, H., Adler-Nissen, J., Process Biochem. 14 (7):6 (1979).
7. Adler-Nissen, J., J. Agric. Food Chem. 24:1090 (1976).
8. Arai, S., Noguchi, M., Yamashita, M., Kato, H., Fujimaki, M., Agric. Biol. Chem. 34:1569 (1970).
9. Joy, L., PAG Bull. 5 (2):7 (1975).
10. Bender, A.E., "The Facts of Food", p. 33. Oxford Univ. Press, London 1975.
11. Hessov, I., Acta Chir. Scand. 143:145 (1977).
12. Adler-Nissen, J., Process Biochem. 12 (6):18 (1977).

13. Adler-Nissen, J., Ann. Nutr. Alim. 32:205 (1978).
14. Adler-Nissen, J., J. Chem. Tech. Biotechnol. 32:- (1982) (in press).
15. Guigoz, Y., Solms, J., Chem. Senses Flavor 2:71 (1976).
16. Pour-El, A., Swenson, T.C., US Patent 3,741,771 (1973).
17. Puski, G., US Patent 3,843,801 (1974).
18. Feldman, J.R., Haas, G.J., Lugay, J.C., Wiener, C.J., US Patent 3,857,966 (1974).
19. Svendsen, I., Carlsberg Res. Commun. 41:237 (1976).
20. Murray, T.K., Baker, B.E., J. Sci. Food Agric. 3:470 (1952).
21. Petritschek, A., Lynen, F., Belitz, H.-D., Lebensm. Wiss. u. -Technol. 5:77 (1972).
22. Clegg, K.M., British Patent 1,338,936 (1973).
23. Clegg, K.M., in "Biochemical Aspects of New Protein Food" (J. Adler-Nissen et al., ed.), p. 109. Pergamon Press, Oxford, 1978.
24. Ishida, K., Yamamoto, A., Nippon Shokuhin Kogyo Gakkaishi 23:524 (1976).
25. Adler-Nissen, J., US Patent 4,100,024 (1978).
26. Adler-Nissen, J., J. Agric. Food Chem. 27:1256 (1979).
27. Lalasidis, G., Sjöberg, L.-B., J. Agric. Food Chem. 26:742 (1978).
28. Iacobucci, G.A., Myers, M.J., Emi, S., Myers, D.V., Proc. IV Int. Congr. Food Sci. Technol., vol. 5, p. 83, Madrid, 1974.
29. Myers, D.V., Ricks, E., Myers, M.J., Wilkinson, M., Iacobucci, G.A., Proc. IV Int. Congr. Food Sci. Technol., vol. 5, p. 96, Madrid, 1974.
30. Roland, J.F., Mattis, D.L. Kiang, S., Alm, W.L., J. Food Sci. 43:1491 (1978).
31. Hempenius, W.L., Chandler, K.L., US Patent 3,876,806 (1975).
32. Fujimaki, M., Arai, S., Yamashita, M., Adv. Chem. Ser. 160:156 (1977).
33. Phillips, R.D., Beuchat, L.R., ACS Symp. Ser. 147:275 (1981).
34. Arai, S., in "The Analysis and Control of Less Desirable Flavors in Foods and Beverages" (G. Charalambous, ed.), p. 133. Academic Press, New York, 1980.
35. Novo Industri A/S, "Alcalase(R) 0.6 L", B-207b, Bagsvaerd, Denmark, 1980.
36. Sejr Olsen, H., Adler-Nissen, J., ACS Symp. Ser. 154:133 (1981).
37. Novo Industri A/S, "Subtilisin-A", IB-169b, Bagsvaerd, Denmark, 1980.
38. Pisecky, J., Dairy Ind. Int. 43 (2):4 (1978).

39. Amerine, M.A., Pangborn, R.M., Roessler, E.B., "Principles of Sensory Evaluation of Food", p. 475. Academic Press, New York, 1965.
40. Roessler, E.B., Pangborn, R.M., Sidel, J.L., Stone, H., J. Food Sci. 43:940 (1978).
41. Cross, H.R., Moen, R., Stanfield, M.S., Food Technol. 32 (7):48 (1978).
42. Kroeze, J.H.A., Chem. Senses Flavor 3:443 (1978).
43. Cogan, U., Moshe, M., Mokady, S., J. Sci. Food Agric. 32:459 (1981).

PROBLEMS OF FAST INDIGENISATION ON THE QUALITY OF FOODS AND BEVERAGES IN NIGERIA

Adegboyega O. Bammeke
M. K. C. Sridhar

Department of Preventive and Social Medicine
University of Ibadan
Ibadan, Nigeria

ABSTRACT

Nigeria, with a population of over 80 million, is fast developing towards self-reliance. Besides the few major food and beverage industries, the majority of rural and urban populations resort to traditional foods such as Amala, Eba, meat stew, fish stew, pounded yam, Jollof rice, snaks, fried Plantains, unpasterized milk, and local fermented drinks. These foods are prepared, stored and sold in open markets on the roadside with little or no environmental sanitation. Some of the foods and beverages consumed by the high socioeconomic groups are imported from other countries in frozen or dried forms. These cold storage units break down at times due to power failures. A study was made on Oyo State on the market sanitation, food poisoning incidents, attitudes of people towards food hygiene, and the quality of a few foods and beverages, viz. fruit, squashes, minerals, beer, peanut butter, jams and other local foods. The study revealed that quality control has been negligible and that some entreprenuers even mislabel their products. The problems are essentially due to ineffective law enforcement agencies, lack of valid statistical figures, unsanitary surroundings, and the sudden banning of importations without the building up of an adequate infra-

ISBN 0-12-169080-6

Table 1. The food preparations commonly consumed together with likely contamination due to inadequate storage facilities.

Food Preparation	Ingredients	Possible Contaminants
Amala meal	Cassava, yam, stew (cow, sheep, poultry, or pig meat), pepper, tomatoes, and mixed vegetables	Salmonella sp. Staphylococcus aureus Clostridium sp.
Eba meal	Gari (cassava preparation), stew	Salmonella sp. Staphylococcus aureus Clostridium
Pounded yam	Yam (boiled or powdered), stew	Salmonella sp. Staphylococcus aureus Clostridium sp.
Fish stew	Fish, pepper, tomatoes, with Eba, Rice, etc.	Bacillus sp. Staphylococcus aureus Vibrio sp.
Rice or Jollof rice	Rice, stews (of meat, fish, etc.)	Staphylococcus aureus Clostridium sp. Bacillus cereus
Milk (unpasteurized or fermented)	---	Salmonella sp.
Snacks	Meat or fish pies, sausages, fried banana, chin-chin	Clostridium perfringes Staphylococcus aureus

structure and personnel to maintain good quality control. This study is of interest to other developing countries moving towards self-reliance.

INTRODUCTION

Nigeria is one of the fast developing countries with a population of over 80 million. While a small percentage in urban centers is westernized in their day-to-day living habits, a majority of the rural population adheres to a traditional way of living. However, there seems to be an apparent inclination to change their living habits. Today, even in rural areas, people like to buy canned or wrapped-up foods, beverages, refrigerated and preserved foods. Most of such food materials were being imported until recently; but not people are trying to start local industries to become self-sufficient. Since 1978 indigenization has become sudden because the Government imposed a ban on the importation of several food items. Though in 1975 there were about 15 breweries, 8 dairies, 11 cannieries, 358 bakeries, and several slaughter slabs in the whole country [1], they have dramatically increased in number, as evidenced from several labels on goods in the supermarkets. An account of the types of foods and beverages in the country, their quality, the problems in keeping high quality standards and certain solutions are presented in this paper.

THE FOOD PREPARATIONS COMMONLY CONSUMED

A variety of raw foods grown or prepared indigenously form the major items of daily food. They include tuberous crops, rice, maize, certain fruits, eggs, fish, and meat. The food preparations, its ingredients, and the possible microbial contamination are given in Table 1. Though most of the preparations are made fresh involving soaking, frying, pounding, or boiling, these could be contaminated on storage for selling. Most of the foods are wrapped-up in leaves that also may be a source of contamination. In addition, a variety of canned or frozen foods and bottled beverages that are sold in supermarkets are also consumed.

THE FOOD- AND WATER-BORNE INFECTIONS COMMONLY REPORTED

Many infections are reported to the Ministry of Health [1] and those that are food or water-borne are given in Table 2. These infections recorded in Oyo State are also common in other

Table 2. The common food- or water-borne infections as reported in Oyo State

Infection/ Disease	1973			1974			1975			1976			1977			1978		
	S	C	D	S	C	D	S	C	D	S	C	D	S	C	D	S	C	D
Cholera	655	122	10	537	172	25	278	14	8	243	15	9	NR	451	33	NR	880	43
Amoebic dysentry	-	84	3	-	126	-	-	51	-	-	22	-	NR	NR	NR	NR	NR	NR
Bacillary dysentry	-	338	5	-	706	2	-	478	2	-	2779	2	NR	NR	NR	NR	NR	NR
Unclassified dysentry	-	110266	11	-	6564	42	-	3563	15	-	3840	21	NR	5656	NR	NR	4984	NR
Poliomyelitis	-	99	-	-	164	-	-	53	-	-	263	-	NR	288	NR	NR	338	NR
Typhoid and paratyphoid	-	80	3	-	56	4	-	44	-	-	54	-	NR	NR	NR	NR	NR	NR
Infectious hepatitis	-	41	-	-	53	-	-	67	3	-	86	-	NR	NR	NR	NR	NR	1
Food poisoning (general)	-	3	-	-	8	-	-	5	-	-	-	-	NR	NR	NR	NR	NR	1

S, Suspected; C, Cases; D, Deaths; NR, No Record.

Table 3. Quality of some of the foods and beverages marketed

Sample	Contaminants
Pineapple drink (canned)*	Staphylococcus and yeast cells present
A sample of beer from a restaurant	A two inch thick sediment at the bottom essentially consisting of phosphate granules
Mixed fruit jam	Some fungal layer on the top
Peanut butter	30 coliforms/gm
Bottle mineral water [6] (7 varieties)	High total plate count at 37°C ranging from 1-10^5/ml. Mostly gram and ve spore bearers, motile vacilli, Diptheroids and yeasts.

*A young girl of two years was admitted to the hospital with food poisoning symptoms after consuming this drink.

states in the country. However, the statistical figures reported may not be complete as many cases will not be reported promptly.

A classical Salmonella food poisoning incident was reported by Ojeniyi [5] during a wedding reception in April 1977. The guests, who were served with sandwiches, chin-chin (a preparation made up of wheat flour, eggs, and sugar), and cakes, developed typical food poisoning syptoms, diarrhea, fever, and vomiting about 12 hours after ingestion of the contaminated food. The death toll was about 10. In the sandwiches, 4,000 million (4×10^9) Salmonella typhimurium were isolated per gram of food. The sandwiches were prepared by a caterer in Lagos (about 120 km from the University of Ibadan), locked-up in the trunk of a car and brought to Ibadan, where the party was held. In the traffic hold-up (which is a normal occurrence) and intense tropical heat, the salmonella multiplied under ideal conditions.

Another case of severe food poisoning was reported [3] from Enugu, where eight people died after consuming cassava meal, which is a traditional staple food in Nigeria.

QUALITY OF SOME FOODS AND BEVERAGES MARKETED

Our laboratory has been engaged over the years in the analysis of foods and water for their chemical and bacteriological qualities [6]. Many established firms and organizations send samples for quality testing. Occasionally as a part of research activities, we also go out to newly established factories and study their manufacturing conditions and the quality of their products. The results of analysis of some of the products (Table 3) reveal that some of them need definite quality control.

FACTORS INFLUENCING THE QUALITY OF FOODS

Quality of Raw Materials

The quality of foods is governed by the type of raw materials used. One of the process materials is water. The quality of water has been very inconsistent due to several factors such as old pipe systems, intermittent supply, inadequate treatment and frequent power failures. The number of unsatisfactory water supplies in Oyo State are given in Table 4. This observation was made over the years on the basis of bacteriological examinations.

Table 4. Percentage of unsatisfactory water supplies in Oyo State

Zone	*Percent unsatisfactory*
Ibadan	*63*
Oshun	*50*
Ife/Ijesha	*85*
Oyo	*83*

Many times vendors or food handlers get from farms materials like spinach, tomatoes, peppers, and other vegetables grown with wastewater [7]. They even wash the materials with gutter water and use it in the food preparations. Under these conditions Sridhar and Pillai [8] observed that a variety of microorganisms including several human pathogens adhere to the crops and will not be removed wven after several washings with clean water. This is a serious problem when such materials are used in salads either uncooked or semicooked.

Unsanitary Trading Environments

Most of the Nigerian markets are built around traders traditionally. They are characterized by extreme congestion, dumping of garbage and other solid wastes at street sides, open drains, outdoor cooking, and an absence of open spaces. Noise, dust, foul odors, flies and rodents, pose problems to the health and well-being of traders and customers [9]. In Ibadan city alone there are 26 such markets providing various food items.

Attitude of Management

When there is a competitive market there will be quality consciousness. Because of the sudden banning of imported foods, many small scale industrialists began producing foods. They have an immediate market. In the absence of adequate quality control inspectors, some of those manufacturers deceptively label their products. In one instance, a bottled orange squash was labeled to be containing 100mg ascorbic acid. Many such industries do not have quality control laboratories; on demand they show some old or spurious reports.

Lack of A Basic Infrastructure

The Nigerian standards Organisations and the Food and Drug Department in the Federal Ministry of Health are two agencies empowered to look into the quality of foods in Nigeria. The food laws are usually introduced on an ad hoc basis to deal with specific problems as they arise; they are not comprehensive. Regulations covering general definitions, rules for inspection, sampling and analysis, labeling and advertising, compositional food standards of all major foods sold, limits on usage and residues in food additives and pesticides, limits for contaminants, and rules for sanitary and hypienic practices in food handling during preparation, storing and marketing are necessary.

Another problem has been a lack of adequate and properly trained field inspectors, backed by well-equipped laboratories operated by well-trained analysts.

DISCUSSION AND CONCLUSIONS

Like many other developing countries, Nigeria is also facing a rapid change from an importing society to a self-sufficient society. In the process, there is a clash between westernization and tradition. As long as the foods were imported, their quality was taken for granted. Only recently has it been realized that all imported foods need not necessarily be of the best quality. There were incidents of food poisonings, substandard foodstuffs, and even outright destruction of under-quality foodstuffs. In addition, there has been considerable wastage of foreign exchange towards this basic food importation. From these points of view, the fast indigenization is a welcome move.

However, this has resulted in the growth of several small scale industries including hotel industries with inadequate knowledge of the quality of their products or services. One cannot get a correct picture of health statistics in relation to quality of foods marketed because many of such cases are not reported to the proper authorities. Even if they are reported, there is no proper follow-up to investigate the incidents. This is due, again, to weaker food and drug control laws, inadequately trained personnel and lack of adequate laboratory facilities within reach.

At present there is little information about training and educational efforts directed towards the food industry or consumers. People also should also come together to form consumer protection societies or agencies to safeguard the well-being of all concerned. When the fast indigenization moves parallel with the building up of a basic infrastructure, it wil be a boon to the country which can boast of the best food and the best health.

REFERENCES

1. Annual Statistical Bulletin. Ministry of Health, Oyo State of Nigeria, 1976.
2. Fox, A. (Ed.) *Hygiene and Food Production.* Churchill Livingstone, Edinburgh and London, 1971.
3. "8 die after cassava meal, Nsukka, Nigeria," *Daily Times Nigeria*, April 10, 1981.
4. Industrial Directory. Federal Ministry of Industries, Lagos, Nigeria, 7th ed., 1975.
5. Ojeniyi, A. A. Clinical epidemiology of Salmonellosis in Ibandan and environs in relation to poultry and man. A thesis submitted for Doctor of Philosophy, University of Ibadan, pg. 45, 1980.
6. Oluwande, P. A., and Ogbechie, M. E. B. C. Bacteriological qualities of bottled mineral water in Nigeria. *J. Med. Pharm. Mkt. 4*, 353-355 (1976).
7. Sridhar, M. K. C. Health aspects of land treatment of sewage. Proceedings of the seminar on the Health of Agricultural workers in Nigeria. African Regional Association of Occupational Health, Dec. 1980. (in press).
8. Sridhar, M. K. C., and Pillai, S. C. Pollution of ground water at Bangalore. Proceedings of the Symposium on Environmental Pollution, Central Public Health Engineering Research Institute, Nagpur, India, pg. 59-80. 1973.
9. Sridhar, M. K. C., Oluwande, P. A., and Okubadejo, A. O. Health hazards and pollution from open drains in a Nigerian city. *Ambio 10*, 29-33 (1981).
10. Thatcher, F. S., Clark, D. S. (eds.). Microorganisms in Foods: Their Significance and Methods of Enumeration. University of Toronto Press, Toronto, Canada, 1973.

FLAVOR NUCLEOTIDES' USAGES IN FOODS

Mitsuru Motono

Food Product Development Laboratories
Central Research Laboratories
Ajinomoto Co., Inc.,
Kawasaki, Japan

INTRODUCTION

In the development of formulated foods, flavoring plays an important role. In flavoring, we use natural extracts such as beef extract, pork extract, chicken extract, fish extract, sea foods extract, vegetable extract and so on. We also use yeast extract, hydrolyzed vegetable proteins and hydrolyzed animal proteins.

Although these flavoring materials contain some amount of flavor nucleotides or glutamic acid in themselves, the additional supplement of flavor nucleotides might be effective in producing formulated foods. In this connection, I will explain the fundamental role and usages of flavor nucleotides in this report.

Savory Substances

There is some experimental data concerned with the relationship between taste and water soluble components of foods, based on chemical analysis and taste tests.[1-8] Fig. 1 shows an example of this relationship.

Recently, the Japanese chemist, Konosu, made crab meat extract from snow crab leg meat, and analyzed amino acids, quaternary ammonium bases, nucleotides and related substances: sugars, organic acids and inorganic ions.

According to the analysis, he formulated the complete synthetic extract with 44 commercially available chemicals.

After confirming the taste similarity of the synthetic extract against the original natural extract through sensory tests, he tried omission tests of all kinds of substances. From his omission tests, he concluded that 12 components, Ala, Arg, Glu, Gly, glycine betaine, AMP, CMP, GMP, Na^+, K^+, Cl^-, and $PO_4{}^{3-}$ were found to contribute more or less toward produc-

ISBN 0-12-169080-6

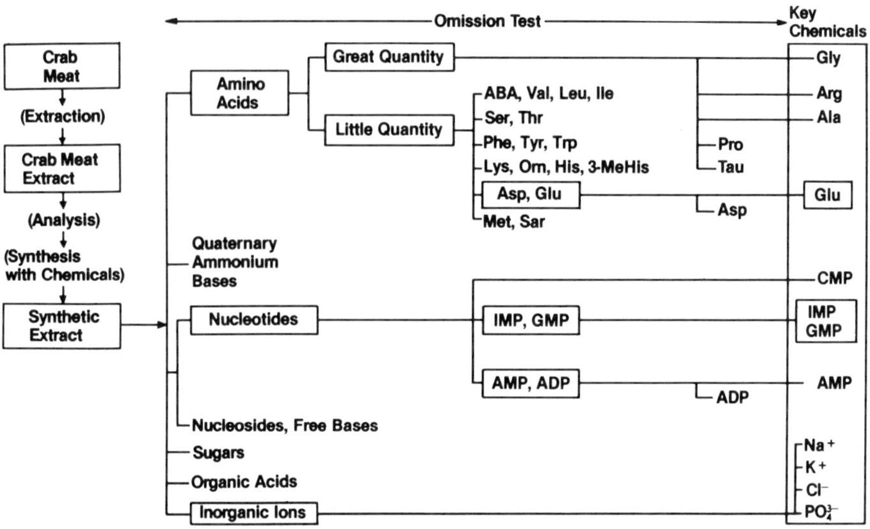

Fig. 1. Summary of Omission Tests

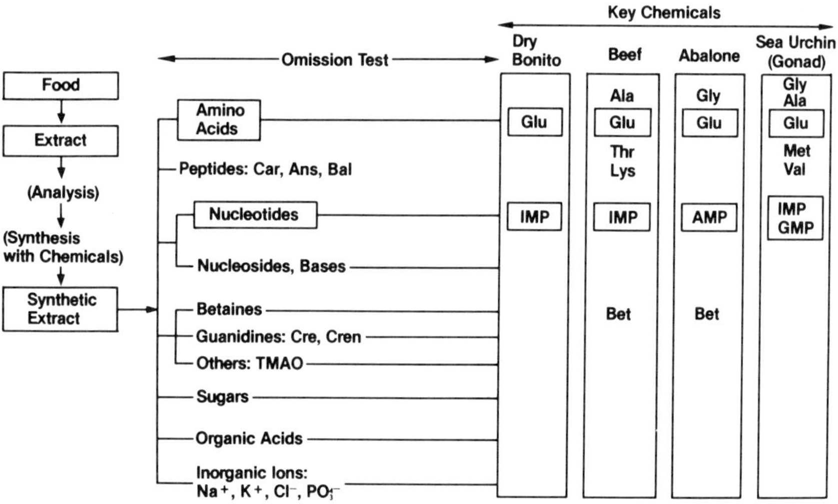

Fig. 2. Constitution of Food Taste Example

ing the characteristic taste of snow crab. The omission of nucleotides and glutamic acid made an especially significant taste difference in his omission tests.

Fig. 2 shows the other results of the omission tests for dried bonito, which is the traditional soup-stock in Japan, beef, abalone and sea urchin (gonad).

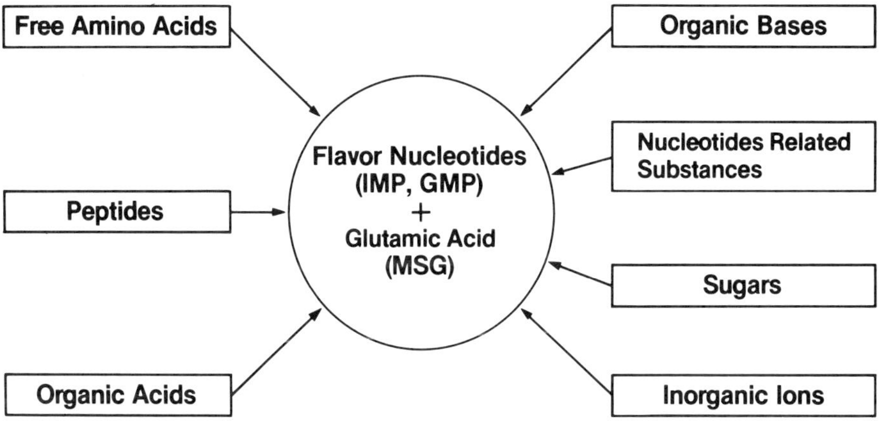

Fig. 3. Constitution of Food Taste

We determined that glutamic acid and IMP are key components for dried bonito; Ala, Glu, Thr, Lys, IMP and Bet. are key components for beef; Gly, Glu, AMP and Bet. are key components for abalone; and gly, Ala, Glu, Met, Val, IMP and AMP, are key components for sea urchin. In all cases, nucleotides and glutlamic acid were key components.

From these studies it is clear that the principal taste components in foods are nucleotides such as disodium inosinate (IMP), disodium guanylate (GMP) and disodium adenylate (AMP), and monosodium glutamate (MSG). (Fig. 3) Each component has at least one of the tastes classified as saltiness, sweetness, sourness, bitterness, or savoriness. However, the taste quality is changed according to the concentrations, pH, and the interaction among them. Consequently, the taste of food is determined by a complex combination of the water soluble components under specific conditions.

Distribution of Nucleotides in Natural Foods

Table 1 shows the nucleotide content in natural foods. From this table, the following facts are observed:

1. Animal foods, such as beef, pork, chicken and dried bonito contain a great amount of IMP.
2. Crustacea and mollusca like squid, lobster, abalone and almost all vegetables contain AMP.
3. The dried mushroom "skiitake," which is a traditional ingredient in Japan and China used in producing soup stock, contains a great amount of GMP.

Table 1. Nucleotide Content in Foods

Material	Nucleotides Content		
	IMP	GMP	AMP
	mg%	mg%	mg%
Beef	163		7.5
Pork	186	3.7	8.6
Chicken	115	2.2	13.1
Whale	326	5.3	2.4
Horse mackerel	323	Ca.O	7.2
Sweet fish	287	Ca.O	8.1
Common sea bass	188	Ca.O	9.5
Pilchard	287	Ca.O	0.8
Black sea bream	421	Ca.O	12.4
Pike markerel	227	Ca.O	7.6
Mackerel	286	Ca.O	6.4
Keta salmon	235	Ca.O	7.8
Tuna	286	Ca.O	5.9
Globefish	287	Ca.O	6.3
Eel	165	Ca.O	20.1
Dried bonito	630 ~ 1310	Ca.O	trace
Squid	0	0	184
Common octopus	0	0	26
Spiny lobster	0	0	82
Hairy Crab	0	0	11
Squilla	26	0	37
Common abalone	0	0	81
Round clam	0	0	98
Common scallop	0	0	116
Short-neck clam	0	0	12
Mushroom, "Shiitake"	0	103	175
Dried-mushroom, "Shiitake"	0	216	321
French mushroom	0	tr	13
Dried-french mushroom	0	tr	190
Mushroom, "Enokidake"	0	32	45
Mushroom, "Matsutake"	0	95	112
Mushroom, "Syorô"	0	9	16
Mushroom, "Hatsutake"	0	85	58

In fact, people have long utilized flavor nucleotides and glutamic acid in cooking without realizing it. For instance, the Japanese still make a soup-stock called "dashi" from dried bonito, which is rich in IMP (0.6 to 1.3g per 100g), along with dried tangle, a kind of sea weed rich in monosodium glutamate (2.3 to 5.4g per 100g). The Chinese, too, still make a soup-stock, "than," from chicken or pork, rich in IMP (0.12 to 0.19g per 100g), and the Europeans still make a soup-stock, "bouillon," from beef rich in IMP (0.16g per 100g) and monosodium glutamate (0.02g per 100g).

Moreover, IMP is produced from ATP, in the following order: ATP → ADP → AMP → IMP. The most savory meat contains the greatest amount of IMP.

The Relationship between Taste and Chemical Structure for Nucleic Acid-Related Substances

The Japanese chemist, Kuninaka,[10] systematically studied the relationship between savory taste and the chemical structure of nucleotide-related substances. He concluded that, as shown in Fig. 4, purine ribonucleotides which have an OH radical on the 6'-carbon of the purine ring and a phosphate ester on the 5'-carbon of ribose moiety have savory taste, while purine ribonucleotides phosphorylated on the 2'-C or 3'-C of the ribose moiety are tasteless. Consequently, IMP, GMP, and XMP (disodium xantylate) have savory tastes. Their savory tastes are strongly enhanced by MSG.

Also, Japanese chemists, Nakao et al.[11] reported that purine deoxyribonucleotides which have an OH radical on the 6'-C of the purine ring and phosphate ester on the 5'-C of the deoxyribose moiety also have savory tastes, but their taste intensities are weaker than those of ribonucleotides. Moreover, Japanese chemists, Honjo et al.[12] reported that the phosphate ester linkage on the 5'-C of the ribose moiety is necessary for imparting savory taste. In addition, the 5'-C phosphate must have both primary and secondary dissociation of OH to exhibit savory taste; if the OH is esterified or amidified, savory taste disappears. On the other hand, synthetic derivatives of nucleotides such as 2-methyl-IMP, 2-ethyl-IMP, 2-N-methyl-GMP, 2-methyl-thio-IMP, and 2-ethyl-thio-IMP are known to have a stronger savory taste in the presence of MSG than normal nucleotides.[13]

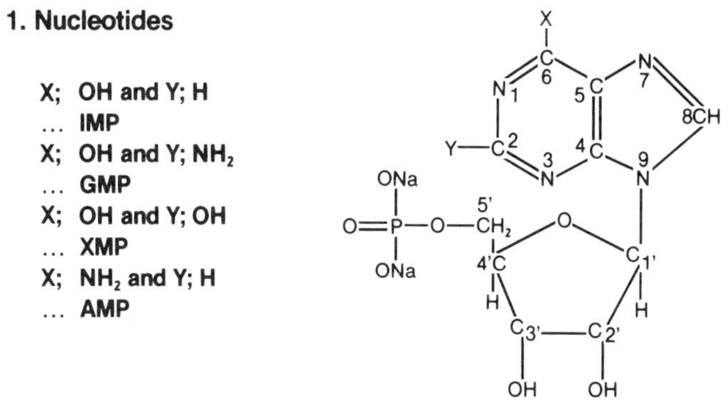

Fig. 4. Chemical Structure of Nucleotides

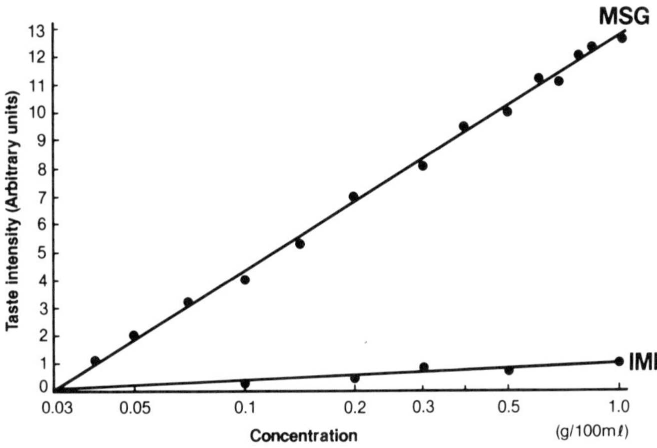

Fig. 5. Relationship between Taste Intensity and Concentration of MSG Solutions and IMP Solutions

Flavor Properties

The threshold values (i.e. the minimal concentrations in water which can be discriminated by taste) of IMP, GMP and MSG are 0.025, 0.0125 and 0.03 g/100mℓ, respectively.

These threshold values are very similar, but the relationship between concentration and intensity of taste, especially at great concentrations, is very different.

As shown in Figure 5, the taste intensity of MSG increases quickly against the logarithm of its concentrations, whereas in the case of nucleotides, the increase is very slight.[14]

Accordingly, while MSG exhibits a marked savory taste by itself, nucleotides do not. The existence of MSG is necessary in order for nucleotides to realize their savory tastes.

Synergism of Savory Taste between Nucleotides and MSG

There is a synergism of savory taste between nucleotides and MSG. Yamaguchi[14] tried to measure this synergism by careful taste tests. She prepared 20 samples composed of four different ratios of IMP and MSG at five levels of concentration, and compared these concentrations at each ratio with five levels of concentrations of MSG alone in order to determine equivalent intensities of savory taste of the mixture and of MSG alone. The taste test was repeated 50 times for each solution, and the results were analyzed by the probit method.

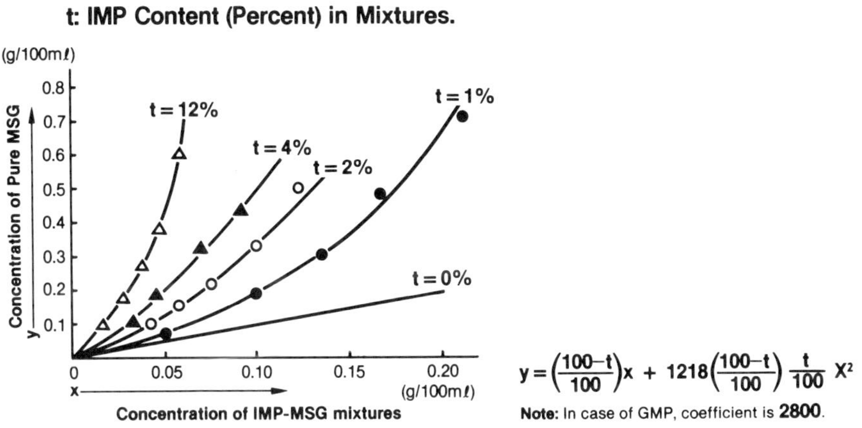

Fig. 6. Taste Equivalency of Mixtures of IMP and MSG to MSG Alone

The equivalent taste intensity between mixtures of IMP and MSG, and MSG alone, are plotted in Figure 6. As shown in Figure 6, taste intensity of mixtures increase exponentially against their concentration, with the degree of synergism depending upon the ratio of IMP to MSG.

From the experiments of taste intensity, she could obtain the equation to determine the adequate ratio of nucleotide to MSG, which can produce the equivalent taste in the case of using MSG alone. When the concentration of MSG in the mixture is u g/100mℓ and that of IMP is v g/100mℓ, and the concentration of MSG alone which shows an equivalent taste intensity is y g/100mℓ, the relationship can be expressed as

$$y = u + 1218\ uv \qquad \text{(Eq. 1)}$$

Similarly, in the case of mixtures of GMP and MSG, replacing v in Eq. 1 by v' g/100mℓ to represent the concentration of GMP and retaining the other variables, the relationship can be expressed as

$$y = u + 2800\ uv' \qquad \text{(Eq. 2)}$$

Comparing Eq. 2 with Eq. 1, the only difference is the coefficients, that of Eq. 2 being 2.3 times greater than that of Eq. 1. This indicates that the synergistic effect of GMP is 2.3 times stronger than that of IMP.

For a mixture of IMP, GMP and MSG, if the concentration of IMP plus GMP is v" g/100mℓ and the ratio of IMP to GMP is P:q, the equation becomes:

$$y = u + \frac{(P+2.3q)}{P+q} \cdot 1218\ uv'' \qquad \text{(Eq. 3)}$$

Yamaguchi[15] determined the variation of taste intensity for mixtures of IMP with MSG at a fixed concentration, 0.05 g/100mℓ of the mixture but with the components present in various ratios from 0:100 to 100:0, using the Gulliksen method. The results are shown in Figure 7.

These experiments were conducted in water or saline solution. Therefore, in the case of the actual formation of processed foods, we have to consider the concentrations of nucleotides and MSG which naturally exist in food ingredients.

Table 2 shows the conversion table of MSG alone into a mixture of MSG and nucleotides in order to obtain an identical flavoring effect on the basis of the equation. If t is 5, 0.051 of the mixture of MSG and IMP can provide a taste identical to 0.2 of MSG alone (i.e. 0.0485 MSG and 0.0025 IMP).

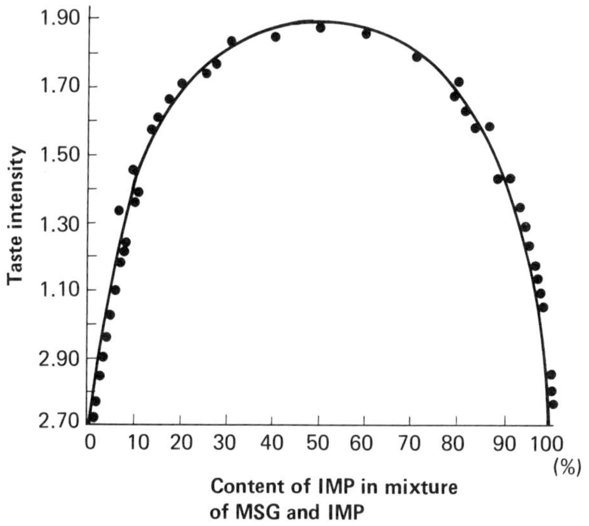

Fig. 7. Consistency between Taste Intensity Determined by the Gulliksen Method and the Calculated (9) Values from Eq. 1

The conversion values in Table 2 were obtained as a result of the use of an aqueous or saline solution. Therefore, the values given are guidelines only.

Table 2. Conversion Table

y = expresses the concentration of MSG alone.
t = expresses the content (%) of nucleotides in the mixture.

y = 0.2 g/100mℓ

t (%)	Concentration of mixture of MSG and nucleotides (X g/100mℓ)		
	In Use of IMP	In Use of GMP	In Use of I + G
1	0.095	0.069	0.051
2	0.074	0.052	0.040
3	0.063	0.044	0.034
4	0.056	0.039	0.030
5	0.051	0.036	0.028
6	0.048	0.033	0.026
7	0.045	0.031	0.024
8	0.043	0.029	0.023
10	0.039	0.027	0.021
12	0.036	0.025	0.020

t: Nucleotide Content in Mixtures

y = 0.60 g/100mℓ

t (%)	Concentration of MSG-Nucleotide Mixtures (X g/100mℓ)		
	In Use of IMP	In Use of GMP	In Use of I + G
1	0.187	0.131	0.152
2	0.140	0.097	0.112
3	0.118	0.081	0.094
4	0.104	0.071	0.083
5	0.095	0.064	0.075
6	0.087	0.059	0.069
7	0.082	0.055	0.065
8	0.077	0.052	0.061
10	0.070	0.047	0.056
12	0.065	0.044	0.052

Physical Properties

In Table 3, the solubility of nucleotides and MSG in water, liquid hydrolyzed vegetable protein and vinegar, all at 20°C, is indicated. IMP shows great solubility in vinegar, while that of MSG is very low. On the other hand, as shown in Table 1 and Fig. 8, GMP becomes gelatinous in acid liquid. Accordingly, IMP has the advantage over GMP of being more soluble and physically stable in liquid hydrolyzed protein or vinegar.

The stability of nucleotides against heat varies with water quantity, pH and temperature. Nucleotides are stable against heat under dry conditions, even when mixed with acid or alkaline substances. On the other hand, their stability against heat in liquids depends on the temperature and the pH.

The stability of IMP in vinegar at 20°C (68°F) during 60 days' storage is shown in Table 4.

The stability of nucleotides against heating at 100°C (212°F) for one and two hours under ordinary pH of foods is shown in Table 5; that of I+G in the canning process of corned beef is shown in Table 6; and that of nucleotides in the frying process of fish is shown in Table 7.

As shown in these tables, nucleotides are almost stable against heat.

Table 3. Soluble Amount of Nucleotides and MSG all in 100g of Water, 100g of Liquid Hydrolyzed Vegetable Protein and 100g of Vinegar, at 20°C

Nucleotide	Water	Liquid H.V.P	Vinegar
IMP	24g	5.6g	30g
GMP	20g	0.5g	12g
MSG	66g	14g	0.9g

Table 4. Stability of IMP in Vinegar at 20°C

Storage periods	0 hr.	18 hrs.	3 days	10 days	30 days	60 days
Survival ratios	100	100	98.1	98.0	96.7	95.6

(The initial concentration of IMP is 0.25%; figures in the table represent percentage of survival ratios.)

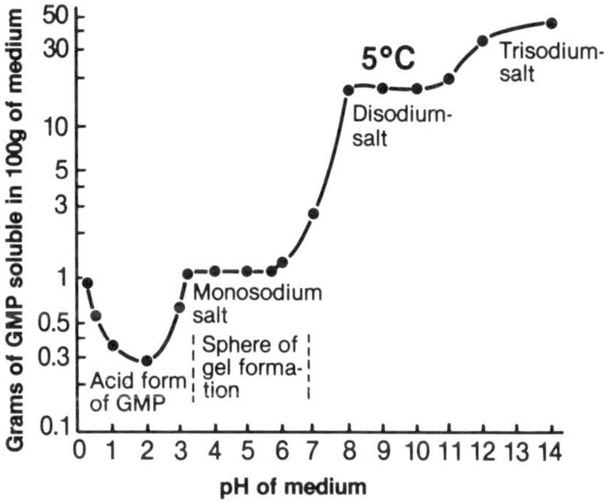

Fig. 8. Solubility of GMP at various pH Values at 5°C

Table 5. Stability of Nucleotides Against Heating at 100°C for One and Two Hours

Nucleotides / Hours / pH	IMP		GMP		I+G	
	1	2	1	2	1	2
5.6	94.9	92.8	88.5	83.5	92.1	86.5
7.0	97.0	92.8	97.2	93.5	97.0	95.6

(The initial concentration of nucleotides in test solutions:
IMP. 0.25%
GMP. 0.11%
I+G. 0.15%
Figures in the table represent percentage of survival ratios.)

Table 6. Stability of I+G in Corned Beef Canning Process

Canning conditions		Survival ratios (%)
110°C	0min.	100
	30min.	98
	60min.	90
	120min.	88
120°C	0min.	100
	30min.	94
	60min.	78

(Dosage of I+G is 100mg in 100g of corned beef.)

Table 7. Stability of Nucleotides in Fish Frying Process

Nucleotides / Frying conditions	IMP	GMP
170°C∼180°C, 3min.	99.7	99.3

(Dosages of nucleotides:
IMP...100mg to 100g of fish
GMP...43mg to 100g of fish
Figures in the table represent percentage of survival ratios.)

Utilization of Nucleotides in Foods

It is an amazing fact that the synergistic effect between nucleotides and MSG has been long established. As I have mentioned already, people have utilized this phenomena in their cooking without being aware of it. And, in fact, we can apply it to the industrial food process.

Table 8 shows an application of IMP to beef-flavored soup as a formulation example. Identical flavors for Soup A and Soup B can be obtained.

Table 9 presents a formulation example of beef extract replacement with IMP. Identical flavors for Form 1 and Form 2 can be obtained.

Table 8. Formulation Example

Soup	A	B
Soup Base*	5.5g	5.5g
MSG	0.6g	0.145g
IMP	—	0.005g
Hot Water	200m*l*	200m*l*

Formulation of Beef-Flavored Soup Base

Hydrolyzed vegetable protein powder	44.7%
Beef extract powder	4.4
Yeast extract powder	4.4
Beef fat	3.3
Salt	20.1
Sugar	13.1
Onion powder	7.2
Celery powder	1.3
Cayenne pepper	0.06
Caramel powder	1.44
	100.00%

From these data, we can reduce the consumption of MSG or beef extract. Whenever we decide the optimum ratio of MSG to nucleotides in processed food, we have to consider the amount of nucleotides and MSG which naturally exists in the other ingredients.

The suggested levels of nucleotides with MSG in processed foods are indicated in Table 10.

Table 9. Formulation Example

Formulations

Ingredients \ Forms	Form 1 (BE, MSG)	Form 2 (1/2BE, IMP, MSG)
"IMP"	none	0.012g
MSG	0.25g	0.14g
Beef extract	0.40g	0.20g
Beef fat	0.55g	0.55g
Vegetable mixture*	0.215g	0.215g
Spice, herbs mixture**	0.046g	0.046g
Salt	0.80g	0.80g
Caramel	none	0.02g
Hot water	100mℓ	100mℓ

* Carrot powder 0.10g, onion powder 0.10g, celery powder 0.015g

** White pepper 0.010g, black pepper 0.010g, garlic powder 0.015g, laurel 0.001g. parsley 0.010g.

Result

Form 2 realized taste identical to that of Form 1.

Table 10. Suggested Usage Levels of Nucleotides

Processed Foods	Levels of MSG with "IMP," "I + G" or "GMP" Required for 10kg		
	MSG (g)	IMP (g)	I + G (g)
Dehydrated soups and gravies (Serving dilution: 1→12~13)	600~800	20~26	12~16
Canned soups and gravies	12~18	0.36~0.54	0.22~0.33
Canned foods (poultry, sausages, ham and fish fillets)	10~20	1.0~1.5	0.60~1.0
Frozen foods (hamburgers)	10~15	0.20~0.40	0.10~0.23
Ketchup	20~30	3.0~6.0	1.8~3.5
Mayonnaise	40~60	2.0~3.0	1.2~1.8
Sauces (Worcestershire type)	30~60	4.0~8.0	2.3~4.7
Vinegar		10~15	7.0~10
Snacks (potato chips, peanuts and crackers)	10~50	0.50~1.0	0.30~0.70
Soy sauce	30~60	4.0~8.0	3.0~5.0
Soup powder for instant noodles (Serving dilution: 1→60~65)	1,000~1,700	50~90	30~60
Dressing	40 ~ 60	2 ~ 3	1.5 ~ 2

CONCLUSION

In the flavoring process, we use many kinds of extracts. Although these flavoring materials contain some amount of nucleotides and MSG, the additional supplement is effective for several reasons. In this report, I have presented an outline of nucleotides as follows:

1. The key components of foods are flavor nucleotides and glutamic acid.
2. All kinds of foods contain MSG and nucleotides such as IMP, GMP and AMP.
3. There is a synergistic effect between MSG and nucleotides.
4. The equation to determine the correct ratio of MSG to nucleotides which produces the taste equivalent to using MSG alone is as follows:

$$y = u + 1218\,uv$$

y: the concentration of MSG alone.
u: the concentration of MSG in the mixture.
v: the concentration of IMP in the mixture.

REFERENCES

1. Y. Hashimoto, F.A.O. Symposium Significance of Fundamental Research in Utilization of Fish (1964)
2. Y. Komata, N. Kosugi, and Y. Ito, Bull. Jap. Soc. Sci. Fisheries, 28, 623 (1962)
3. Y. Komata, H. Eguchi, Ibid., 28, 631 (1962)
4. Y. Komata, A. Mukai, and Y. Okada, Ibid., 28, 747 (1962)
5. Y. Komata, Ibid., 30, 749 (1964)
6. T. Fujita and Y. Hashimoto, Ibid., 26, 907 (1960)
7. A.E. Bender, T. Wood, and J.A. Palgrave, J. Sci., Food Ar., 9, 812 (1958)
8. T. Wood and A.E. Bender, Biochem. J., 67, 366 (1957)
9. S. Konoso, J. Food Sci., 46(2) 479-483 & 493 (1981)
10. A Kuninak, Nippon Nogeikagaku Kaishi (Japanese), 34, 489 (1960)
11. Y, Nakao and K. Ogata, Second Kanto Branch Congress Agr. Chem. Soc. Japan, 1960 Abstracts, P.A.
12. M. Hanjo, K. Imai et al. Annual Congress Agr. Chem. Soc. Japan, 1963 Abstracts, P.40.
13. S. Yamaguchi, et al. Agr. Bis. Chem., 32, 797 (1968)
14. S. Yamaguchi, J. Food, Sci., 32, 473 (1967)
15. S. Ikeda, H. Furukawa, and S. Yamaguchi, Hinshitsu Kanri (Japanese), 13, 768 (1962)
16. S. Gulliksen, Psychrometrica, 21, 125 (1907)

MANUFACTURE, USE, AND NUTRITIONAL ASPECTS OF 90% HIGH FRUCTOSE CORN SWEETENERS

L. Steven Young

Archer Daniels Midland Co.
Decatur, IL

John E. Long

Archer Daniels Midland Co.
Cedar Rapids, IA

Few food ingredients in contemporary society have had the industrial impact as have high fructose corn sweeteners (HFCS). Few food ingredients in contemporary society have been the beneficiary of modern, efficient, integrated production processes, of versatility in use and of excellence in quality as have high fructose corn sweeteners. These are the established patterns of high fructose corn sweeteners. Patterns that allow replacement of 50-100% of sucrose improved product quality, and significantly reduced ingredient costs in processed foods.

By 1985, corn-derived sweeteners (regular corn syrup and HFCS) will command over 40% of the nutritive sweetener market (Figures 1 and 2). Technically 50% of the sucrose used in the U.S. could be replaced with corn-derived sweeteners. In addition, the superior sweetness and potential metabolic benefits of the simple sugar, fructose, could further market demand. These are the emerging patterns of high fructose corn sweeteners.

Why use high fructose corn sweeteners? Because HFCS products are relatively new (10-15 years), modern facilities are used in production. Labor is minimized and manufacturing costs are lower than for sucrose. Products are readily available and distributed bulk to all parts of the continental United States. For these reasons, the use of HFCS is most often more economical than when using sucrose. Finished product quality is maintained or improved. The excitement and potential of new market opportunities also is of interest. This, particularly with the commercial availability of 90% HFCS for reduced calorie products and products formulated for specific metabolic benefits.

ISBN 0-12-169080-6

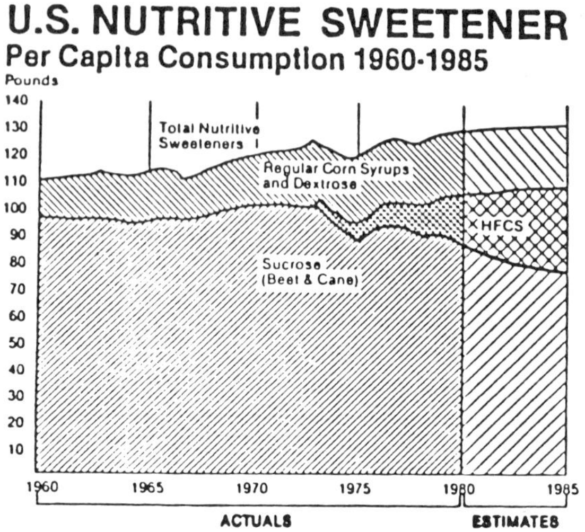

Fig. 1. Per Capita Consumption of Nutritive Sweeteners in the United States, 1960-1985 (USDA Data with Amstar [1980] Projections).

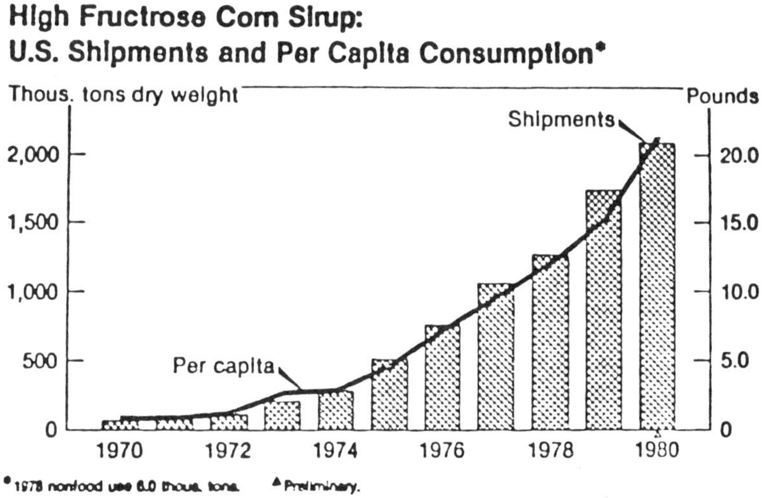

Fig. 2. United States Shipments and Consumption of High Fructose Corn Syrup (USDA, 1980)

I. MANUFACTURE, COMPOSITION, AND PROPERTIES

From one of America's most abundant crops, Number 2 Yellow Dent Corn, the corn wet-milling process yields a variety of products for feed, food, and fuel (Fig. 3, Table 1). Gluten feed (21% protein for feed), gluten meal (60% protein for feed), defatted germ (for feed), and oil (for

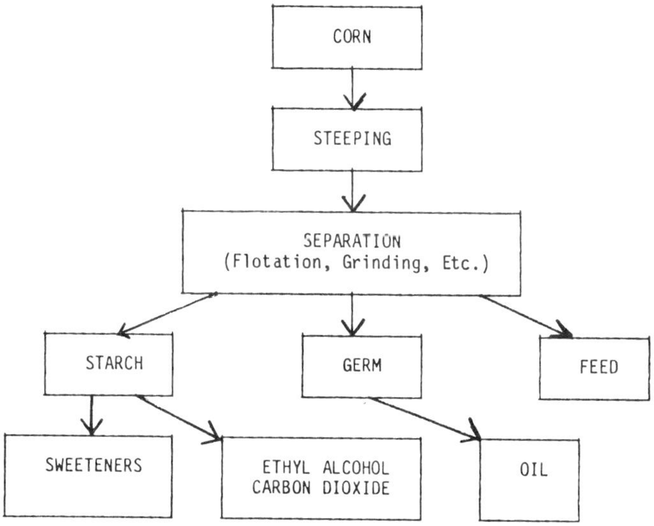

Fig. 3 Steps in the Wet-Milling of Corn

TABLE 1. Products Derived form the Wet-Milling of One Bushel of Corn[a]

Product	Pounds Dry Solids
Feed, 21% Protein	9.2
Gluten Meal, 60% Protein	2.7
Germ[b]	3.5
Starch	31.5

[a]Bushel of corn at 15% moisture weighs 56 pounds (47.32 pounds dry solids).
[b]45-55% corn oil.

margarine, salad oils, shortenings, etc.) constitute only one-third of the dry solids of a bushel of corn. By themselves, these "co-products" can return a significant portion of the cost of a bushel of corn. This reduces the cost of starch from which sweetener products are manufactured. Starch can be converted to make: dextrose (glucose), regular corn syrup and HFCS. These provide sweetness and beneficial functionality to food products. Although somewhat sweet, dextrose and regular corn syrups lack sufficient sweetness to adequately replace sucrose in some foods. The intense sweetness of high fructose syrups, in particular 90% HFCS, offer significant new market opportunities.

The rapid development of new enzyme technologies to isomerize dextrose (70-80% as sweet as sucrose) to fructose (20-60% sweeter than sucrose) has enabled the design of systems to manufacture products as sweet as, or sweeter than, sucrose. These are the high fructose corn sweeteners. Figures 4 and 5 outline the processes by which 42, 55, and 90% high fructose corn sweeteners are produced. Compositional comparisons to commercially available sucrose products and regular corn syrups are given in Table 2. HFCS products are delivered at relatively high solids (71-77% dry solids) and low pH's (3-4), are essentially all monosaccharide in composition, have high osmosic pressures and are microbiologically stable. Due to their monosaccharide nature, all HFCS products, compared to sucrose or regular corn syrups, will depress freezing points, increase boiling points, reduce vapor pressure and increase osmotic pressures in foods. They are hygroscopic and tend to pick up or retain moisture in formulated foods. The greater the percentage of free fructose, the greater this tendency. Fructose and dextrose are reducing sugars and will participate in Maillard-type (nonenzymatic) browning. Enhancement of flavors, particularly fruit flavors, is a typical characteristic of HFCS. As monosaccharides, HFCS products are almost totally fermentable. Storage and handling characteristics allow HFCS to be handled through properly designed liquid handling systems.

Sweetness is, of course, the most important general property of the high fructose corn sweeteners. Relative sweetnesses of various nutritive carbohydrates are compared in Table 3.

The actual sweetener system of choice depends on desired sweetness, functionality, available storage facilities, product(s) to be formulated, economics, and good judgment. Major market demands for HFCS are given in Table 4.

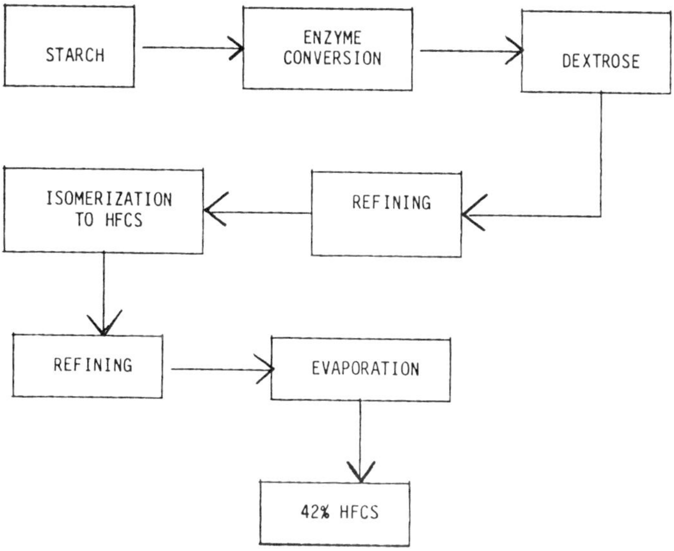

Fig. 4. Steps in the Production of 42% High Fructose Corn Sweetener.

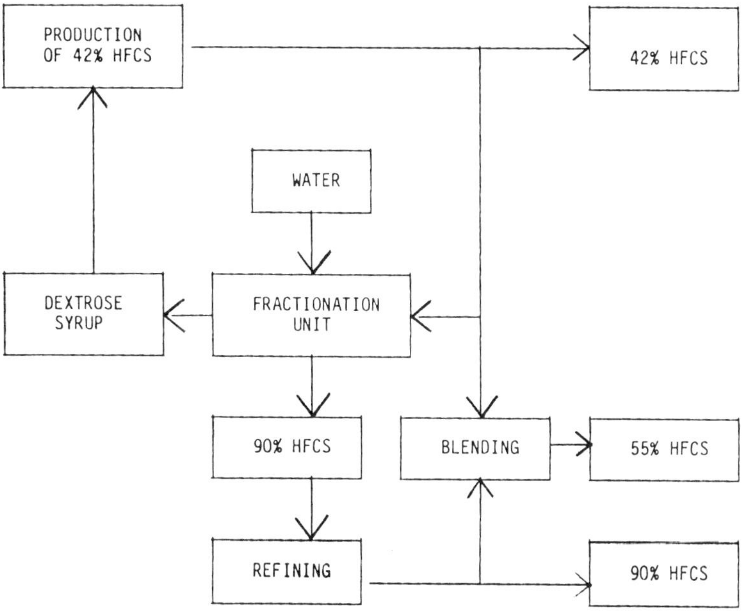

Fig. 5. Steps in the Production of 55 and 90% High Fructose Corn Sweetener.

TABLE 2. Carbohydrate Profile of Various Nutritive Sweeteners

Sweetener	DP_1[a]		DP_2		DP_3	DP_4 and Higher (%)
	Fructose (%)	Dextrose (%)	Sucrose (%)	Maltose (%)	Triose (%)	
Sucrose			100			
Medium Invert	25	27	46	2[b]		
Total Invert	45	48	3	4[b]		
Dextrose		100				
42 D.E. CSU		20		14	12	54
42 D.E. High Maltose CSU		8		40	15	37
62 D.E. CSU		39		31	7	23
42% HFCS	42	52		6[b]		
55% HFCS	55	40		5[b]		
90% HFCS	90	9		1[b]		

[a]DP = Degree of polymerization.
[b]DP_2 and higher saccharides.

TABLE 3. Relative Sweetness of Nutritive Sweeteners

Sweetener	Sweetness Relative to Sucrose
Sucrose	1
HFCS (90%)	1.2-1.6
HFCS (42%)	1
Medium Invert	1+
HFCS (55%)	1+
Dextrose	0.7
Maltose	0.5
Lactose	0.2
Sorbitol	0.5
Xylitol	1

TABLE 4. Major Markets for HFCS

Market Use	Portion of Total Supply %
Beverages	40
Baking	24
Canning	14
Desserts and ice cream	8
Jams and preserves	4
Other	10

Of special interest is 90% HFCS (Table 2). The commercial availability and unique functionalities of 90% HFCS have created new application frontiers in some standard food formulations and, most significantly, in "light" or reduced calorie products. Ninety percent HFCS offers superior sweetness (Table 2) and significant advantages (Table 5) in formulating calorically reduced foods. Because 90% HFCS is normally sweeter than sucrose, less can be used to generate products with significant calorie reductions. Care must be taken in that sweetness varies from system to system. There is no substitution for in-product testing. The product is the most stable of all liquid sweeteners (77% dry solids at pH 3-4), resists crystallization, and thus can be easily handled at ambient temperatures. In addition, 90% HFCS further enhances flavors, is more hygroscopic, and more readily participates in Maillard-type browning, and more soluble than either 42% or 55% HFCS. Like other HFCS products, 90% HFCS depresses freezing points compared to higher molecular weight species (e.g., sucrose and higher saccharides). All properties should be taken into account when reformulating existing foods or formulating new concepts.

TABLE 5. Functional Properties of 90% HFCS

Sweetness
Flavor enhancement
Humectancy
Solids
Solubility/fluidity
Microbiological stability
Handling

II. APPLICATIONS OF 90% HIGH FRUCTOSE CORN SWEETENER

Table 6 gives calorie reductions possible when using 90% HFCS (Young, 1981, 1979).

TABLE 6. Reduced Calorie Prototype Guidelines Using 90% HFCS

Products	Calorie Reduction
Dairy products-fluid/frozen	At least 1/3
Frozen "capped" fruit-bulk or consumer pack	At least 1/3
Soft drinks/still drinks	Up to 50%
Jam/jelly-type spread	At least 1/3
Flavored syrup concentrates	At least 1/3
Canned fruit	Up to 50%
Table Syrups	40-50%
Confections	Depends on type
Salad dressings	45-50%
Baked goods	Depends on type
Dry cereal	At least 1/3
Table top sweetener	Up to 50%[a]

[a]Sweeten to taste.

Dairy products - it is relatively easy to develop reduced calorie fluid concepts (i.e., fruit drinks and non-fat chocolate drinks). However, care must be taken when formulating low solids, reduced calorie frozen products. Total reformulations are normally required. Fat is reduced as are total solids. Sweetener reduction only partially aids total calorie reduction. Such adjustments will result in predictable loss of body, texture, mouthfeel, and stability to heat shock. However, careful reformulation can result in products with exceptonal performance and quality.

Frozen "capped" fruit (bulk or consumer packs) - the same property that effects frozen dessert performance (i.e., freezing point depression) also results in superior quality frozen fruit. One-third to one-half less sweetener can be used. Calorie reductions, while enhancing flavor and maintaining color and fruit integrity, are possible.

Soft drink/still drinks - up to 50% reduction in calories is possible given appropriate systems. it sometimes is advisable to increase flavor and/or reduce acid to maximize sweetnes. The use of gums or other stabilizers are advantageous in increasing viscosity and body lost by reducing sweetener solids. This helps to maintain perceived sweetness and total flavor profile without significantly increasing calories or cost.

Jam/jelly-type spreads - a 30-50% calorie reduction is possible simply by reducing sweetener solids. Low solids products are not standard and will require different gelling systems (e.g., low methoxyl pectin and/or other gums) for performance.

Flavored syrup concentrates - the superior sweetness and solubility of 90% HFCS allows formulation of beverage concentrates with higher mix ratios than previously attainable with sucrose or medium invert. Quality can be maintained while improving yields and storage stability, reducing labor, transportation, and sweetener costs, and reducing calories.

Canned fruit - since the bulk of calories in these items is a result of added sweetener, reduced calorie systems are possible by using less 90% HFCS. Fifty percent calorie reductions still yield high quality fruit.

Table syrups - when reducing solids to achieve a reduction in calories, loss of mouthfeel, body and viscosity must be compensated by using gums or other stabilizers. This normally is quite simple to accomplish.

Confections - depending on type, reduced calorie products are possible. Again, as other components (i.e., fat and other sweeteners) are formulated out, functionality must be replaced. Care must be taken so that sticky and/or gummy products do not result. The humectant characteristics of 90% HFCS often limit its usage.

Salad dressings - sweeteners compose only a small percentage of the solids of both spoonable nd pourable types. The 90% HFCS can only help reduce calorie load. Body, mouthfeel, viscosity, and "cling" properties must be replaced as oil is decreased.

Baked goods - development of reduced calorie baked goods is possible. Functionality in many instances is critical and difficult to replace.

Dry cereals - Reduced sweetener cereals are possible. The hygroscopic nature of 90% HFCS limits use in only compatible products. Reduced sweetener in the body of the cereal can only h elp reduce calories. Excessive moisture pickup can yield soft piece identity. As a coating, the 90% HFCS does not crystallize as does sucrose and thus may yield sticky and gummy pieces.

Table top sweetener - easy to dispense in squeeze bottles, spouted containers, or portion-controlled packets, 90% HFCS permits ambient temperature distribution and use. Up to 50% less sweetener can be used for cereals, beverages, canned fruits and other products. The sweetener is easy to apply to home recipes given slight modification for added water and/or perceived sweetness. It is recommended to always sweeten to taste.

III. NUTRITION ISSUES PERTANING TO 90% HFCS

Recent research has been done evaluating nutritional aspects of the sugar fructose. However, the commercial realities of 90% HFCS have given new emphasis to nutrition research, both basic and applied. Numerous topics have been, and are being, investigated.

Work at the University of Iowa Medical School with normal and diabetic patients has demonstrated the presence of a so-called "disaccharide effect" in humans (Thompson, et.al., 1979). Previously, this effect was known only to occur in laboratory animals. Significant reductions in integrated (across 24 hours) blood triglycerides were noted in patients consuming monosaccharides (fructose and glucose) versus equal amounts of fructose and glucose as the disaccharide sucrose. The effect was present in normal and diabetic subjects. Although the mechanism is still unknown, investigators now seem intent on establishing a definite advantage in consuming monosaccharides (i.e., glucose and/or fructose) in comparison to disaccharides (i.e., sucrose).

A. General Metabolic Aspects of Fructose

The general metabolic effects of fructose consumption are fairly well documented (Crapo and Olefsky, 1980; Olefsky and Crapo, 1980; Brunzell, 1978; Forester, 1974). Fructose is more slowly absorbed into the portal vein by facillitated diffusion in the gastrointestinal tract than glucose. Although some fructose is converted to glucose and/or lactate during this process, 80-90% of absorbed fructose remains unmetabolized. The half life of fructose in the portal vein is extremely short. Fructose is rapidly and primarily metabolized in the liver. Although other tissues utilize fructose, only minor elevations of plasma fructose are noted (Crapo and Olefsky, 1980). Thus, the rate limiting step in fructose metabolism is its rate of absorption into the blood. The initial steps of fructose metabolism are insulin independent and help to spare insulin. The metabolic fate of fructose and other simple carbohydrates in the liver are outlined in Figure 6 (Kovisto, 1978).

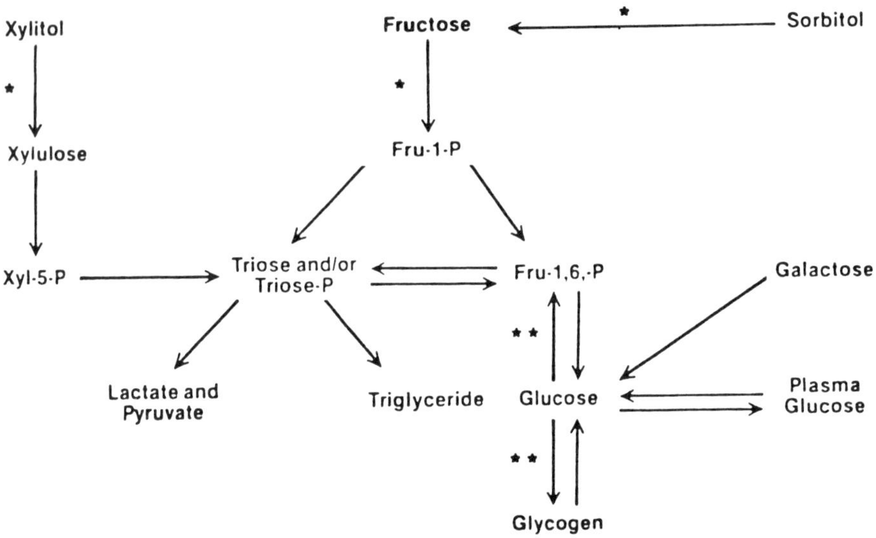

Figure 6. Metabolism of Simple Carbohydrates in the Liver

Fructose is rapidly phosphorylated to fructose-1-phosphate. Failure to phosphorylate fructose results in essential fructosuria. The liver simply does not metabolize dietary fructose. This condition is rare and benign, and subjects are asymtomatic. However, failure to reduce fructose-1-phosphate to D-glyceraldehyde and dihydroxyacetone phosphate results in more serious hereditary fructose intolerance. Accumulation of fructose-1-phosphate and reduction in free inorganic phosphate can result in hypoglycemia, vomiting, and other symptoms. The condition is more serious in small children, and careful monitoring is required. Individuals become aware early in life that dietary exclusion of all forms of sweets (e.g., glucose, fructose, sucrose, fruits, and other foods) is necessary. Normally, subjects also acquire a reversion to sweet foods in general. Treatment is simple exclusion of these products from the diet (Crapo and Olefsky, 1980; Hue, 1974).

The fate of fructose-1-phosphate depends on the metabolic state of the organism. Synthesis of glucose or triglyceride is possible as is the generation of energy via the glycolytic pathway. In normal individuals, glycogen is formed from glucose with minor release of glucose to the blood. Given fructose's slow absorption rate, only slight increases in blood glucose levels are noted after fructose ingestion. Maintenance of blood glucose levels results. This has been found of value to normal subjects with impared glucose tolerance (i.e., hypoglycemics) and overt non-insulin dependent (Type II) diabetes mellitus.

Hypoglycemia - Reactive hypoglycemia is relatively difficult to diagnose. Normal individuals do experience "low blood sugar" after consumption of sucrose and/or glucose. However, many do not exhibit true reactive hypoglycemia. Inordinately low blood glucose levels (less than 60 mg %) must be documented as well as the presence of typical symptoms. Treatment is dietary and often difficult to follow. Crapo and Olefsky (1980) note that individuals usually abuse dietary regimes and consume large amounts of sucrose. This practice, and the tendency to be reactive, produces the typical symptomatic disorder. Bohannon, et.al. (1980) and Crapo and Olefsky (1980) report that when fructose is ingested, a definite maintenance in blood glucose results. In patients with reactive hypoglycemia, blood glucose levels fail to fall below 65 mg % after three to five hours after ingestion. Clinical symptoms were not noted.

Diabetes - Diabetics with severe insulin deficiency are unable to take advantage of the metabolic effects of fructose. Liver function is impared so that glucose formed from fructose is not stored as glycogen. The ability of fructose to maintain post prandial blood glucose levels and to spare insulin has been found to be of potential value in the diet of non-insulin dependent diabetics. The concept has been supported by several major studies (Crapo and Olefsky, 1980, Bohannen, et.al., 1980). It is known that diabetics crave sweets and maintain dietary regimes better if allowed sweets (Brunzell, 1978), thus, use of fructose in these diets seems natural. The Committee on Food and Nutrition of the American Diabetes Association has endorsed the position that oral use of limited amounts (less than 75 g per day) of fructose is acceptable and may ultimately provide advantages in dietary management (Crapo and Olefsky, 1980; Olefsky and Crapo, 1980). Therefore, fructose appears to be an acceptable nutritive sweetener for mild or moderately well controlled diabetic patients. Long term studies are being done to confirm this position.

Weight Control/Obesity - Theoretically, the use of fructose to reduce calorie consumption and to maintain blood glucose levels to satiate appetite could aid in the treatment of obesity. Crapo and Olefsky (1980) note that there exists no convincing evidence to support the use of fructose in weight control programs. However, Wiesner, et.al. (1979) have reported that the use of fructose in a diet plan is a safe and effective way of promoting weight loss. These investigations showed an average of 7.6% loss of body weight among twenty-one obese patients across four weeks. No abnormal blood analyses were noted. There was unanimous satisfaction with the dietary plan, and 95% of those participating expressed no feeling of hunger.

Dental Caries - Sucrose is a requirement for the microbial formation of the polysaccharide, glucan, responsible for the formation of dental placque. Placque consists of glucan and adhered micro-organisms. Localized production of lactic acid results in carie occurance. Studies indicate significant differences in placque formation between sucrose and the monosaccharides, glucose and fructose (Frostell, et.al., 1967). Work in Europe (Markinen and Scheinin, 1971) further notes reduction in caries by 25% when individuals were fed fructose sweetened diets compared to sucrose diets. The simple sugars sucrose, glucose, and fructose are fermentable and, thus, acidogenic. Some sources indicate little or no difference in the cariogenic characteristic of sucrose, glucose, and fructose (Alfano, 1981; Schachiel, 1980). Other studies have noted fructose significantly reduces placque formation and carie occurance in laboratory animals (Frostell, 1967; Scheinin and Markinen, 1971) and polysaccharide production in microflora organisms (Gibbons and Baghart, 1967). Complicating the issues are factors such as food form (solid versus liquid), oral microflora, food composition, and frequency of eating (Alfano, 1981).

Alcohol Metabolism - The metabolism of fructose in the liver indirectly accelerates the oxidation of ethanol and relieves ethanol-induced hypoglycemia (Yeikalari, et.al., 1972). Ethanol oxidation reduces NAD+ to NADH via the activity of alcohol dehydrogenase. Tygstrup, et.al., (1965), suggested that fructose, as a source of glyceraldehyde, allows the oxidation of NADH to NAD+, which in turn allows the oxidation of ethanol to proceed unabated. Otherwise, serious inhibition of glucose production results in ethanol-induced hypoglycemia. The conversion of fructose to glucose seems to be uninhibited during ethanol oxidation and, thus, is effective in restoring blood glucose levels.

Athletic Performance - Atheltes that participate in carbohydrate-loading (i.e., glycogen) dietary programs may find advantages in consuming fructose in these diets. Theoretically, the metabolic effects of fructose help to spare glycogen (and insulin) and maintain blood glucose levels. During or after competition, water repletion is essential (Kostil and Saltin, 1974). High soluble solids restrict water absorption rates. The sweetness of fructose helps to minimize soluble solids and maximize water absorption rates when consuming athletic beverages. Performance improvements, based on the metabolic attributes of fructose, are being investigated (MacMurray, 1981).

The emerging patterns are simply the established patterns of process development, product applications, and significant new metabolic benefits. The formulation of new products with 90% HFCS offers potential for foods with reduced calories and for special dietary markets.

REFERENCES

Alfano, M. C., Cereal Foods World 26(1) 5 (1981).

Amstar Corp., "Overview of U.S. and Raw Sugar Markets", Economic Research and Development Section (1980).

Bohannen, N. V., Karam, J. H., Forsham, P. H., J. American Dietetic Association 76(6) 555 (1980).

Crapo, P. A., and Olefsky, J. M., Nutrition Today 15(4) 10 (1980).

Forester, H., Siple, H. L., and McNutt, K. W., ets. Sugars in Nutrition, p. 259, Academic Press, New York (1974).

Frostell, G., Keyes, P. H., and Larson, R. H., J. Nutrition 93(1) 65 (1967).

Gibbons, R. J., and Banghart, S. B., Arch Oral Biochem 12 11 (1967).

Hue, L., in Siple, H. L., and McNutt, K. W., eds., Sugars in Nutrition, p. 357, Academic Press, New York (1974).

Kostil, D. C., and Saltin, B., J. Applied Physiology 37(5) 679 (1974).

Kovisto, V. A., Diabetes Care 1 241 (1978).

MacMurray, R., University of North Carolina, Durham, NC, Personnel Comjunication (1981).

Markinen, K. K., and Scheinin, A., International Dental J. 21 23 (1971).

Olefsky, J. M., and Crapo, P., Diabetes Care 3(2) 390 (1980).

Schachtele, C. F., Contemporary Nutrition 5(7 and 8) (1980).

Scheinin, A., and Makinen, K. K., International Dental J. 21 302 (1971).

Thompson, R. G., Hayford, J. T., and Hendrix, J. A., Science 206 838 (1979).

Tygstrup, N., Winkler, K., and Lundquist, F., J. Clinical Investigation 44 817 (1965).

U.S.D.A., Sugar and Sweetener Report 5(7) 19 (1980).

Young, L. S., Food Prod. Devel., 13(8) 34 (1979).

Young, L. S., Proceedings Stokely-Van Camp Symposium, University of Tennessee, Knoxville, TN, 1981.

Ylikahari, R. H., Kahonen, M. T., and Hassinen, Il, Acta Medica Scandinavica Supplement 542 141 (1972).

Wiesner, T., Walsh, D., Doty, N., and Olerod, C., Current Therapeutic Research, 26(6) 885 (1979).

Functionality of Corn-Derived Sweeteners in Formulated Foods

L. Mark Hanover

A. E. Staley Manufacturing Company
Decatur, Illinois

I. INTRODUCTION

The consumption of sweeteners worldwide has gradually increased over the last one hundred years or so to an average of approximately 46.5 lbs. per capita per year (1978 USDA figures). In the United States, although not the highest in the world, the figure approaches 130 lbs. per capita per year. This figure represents the pounds of refined cane and beet sugar, plus the pounds (dry solids) all corn derived sweeteners and also includes the amount of honey and other edible syrups. In 1970, per capita consumption of refined sugar in the United States was 101.8 lbs., and consumption of corn derived sweeteners (solids) was 19.3 lb. Estimated USDA figures for 1980 place these values at 84.9 lbs. for sugar and 41.1 lbs. of corn derived sweetener solids. The corn derived sweeteners include regular corn syrups, dextrose and high fructose corn syrup.

Corn derived corn sweeteners have become a significant factor when one considers choices for a sweetener ingredient to be utilized in a formulated food. Both sucrose and corn derived sweeteners are classified as "nutritive," meaning they consist of simple digestible carbohydrates that are metabolized by the body and contribute to the caloric value of the food. More specific examples of nutritive sweeteners include fructose (fruit sugar), glucose (dextrose or corn sugar), lactose (milk sugar), maltose and the sugar alcohols such as sorbitol and mannitol. Non-nutritive sweeteners such as saccharin and cyclamates are not metabolized, their usage is, therefore, most popular in the low cal, dietetic food area.

CHEMISTRY OF FOODS AND BEVERAGES:
RECENT DEVELOPMENTS

ISBN 0-12-169080-6

II. MANUFACTURE

Corn syrups, both regular and the newer high fructose products, are actually blends of simple sugars - saccharides - with higher molecular weight multiples of these simple sugars - polysaccharides. The primary simple sugar in conventional (or "regular") corn syrups is dextrose. Dextrose (D-glucose), or the anhydro glucose unit, is the ultimate end product of starch hydrolysis. The process by which corn starch is converted to corn syrups is given in Figure I. This is referred to as the wet milling process because it utilizes water to facilitate the various separation steps. As the diagram shows, starch to syrup conversion is done after the separation of the starch fraction from the corn kernel.

Figure II shows the actual methods by which starch may be "converted" to corn syrups. During these processes, the term D.E. first becomes important. The abbreviation D.E. stands for "dextrose equivalent," which is defined as the measurement of the total reducing sugars in the syrup calculated as dextrose. The values are expressed in relationship to dextrose being 100 D.E. Also, D.E. values are calculated on a dry solids basis (dsb). The D.E. of a syrup can be approximated by calculation from its saccharide distribution when it is known using pre-determined D.E. values for each dextrose polymer present. Or, it can be more accurately determined by employing one of the standard reducing sugar tests such as copper sulfate in Fehling's Solution.

As shown in Figure II, there are three commercially used methods of hydrolyzing corn starch to corn syrups. All methods are not and in fact, cannot, be used to make all types of syrups, but rather each method is used to produce products with individually different saccharide distributions. The theory behind each method is basically the same. In acid catalyzed conversion systems, the bonds linking the anhydroglucose units are broken at random. The time and temperature conditions of the reaction determine the ultimate lengths and distribution of the dextrose polymers. With enzymes, hydrolysis is more specific to some extent depending on the enzyme used. The α-amylase enzyme produces mostly dextrose polymers of intermediate to

FIGURE I. CORN WET MILLING PROCESS

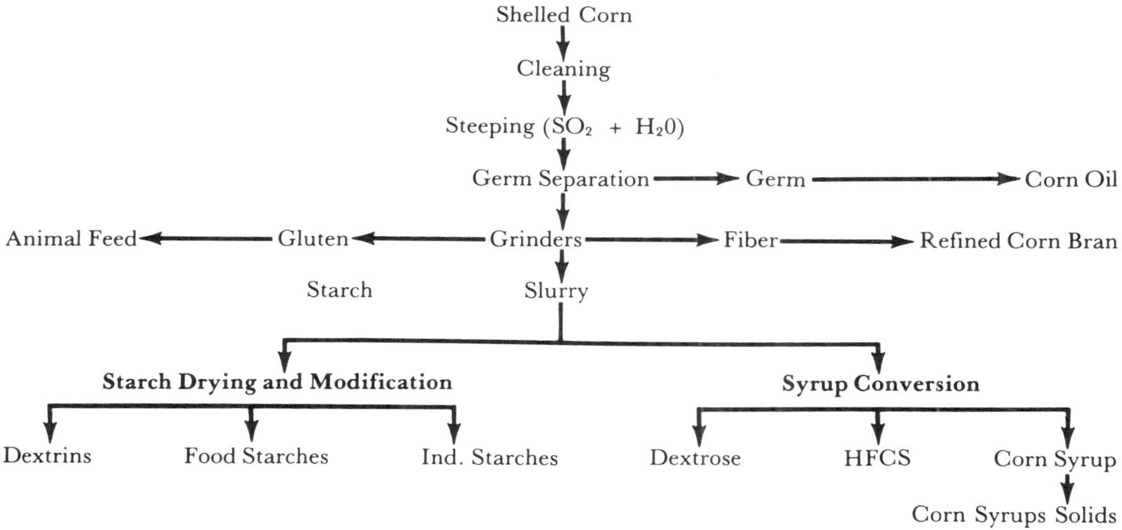

FIGURE II. CONVERSION OF STARCH TO SYRUPS

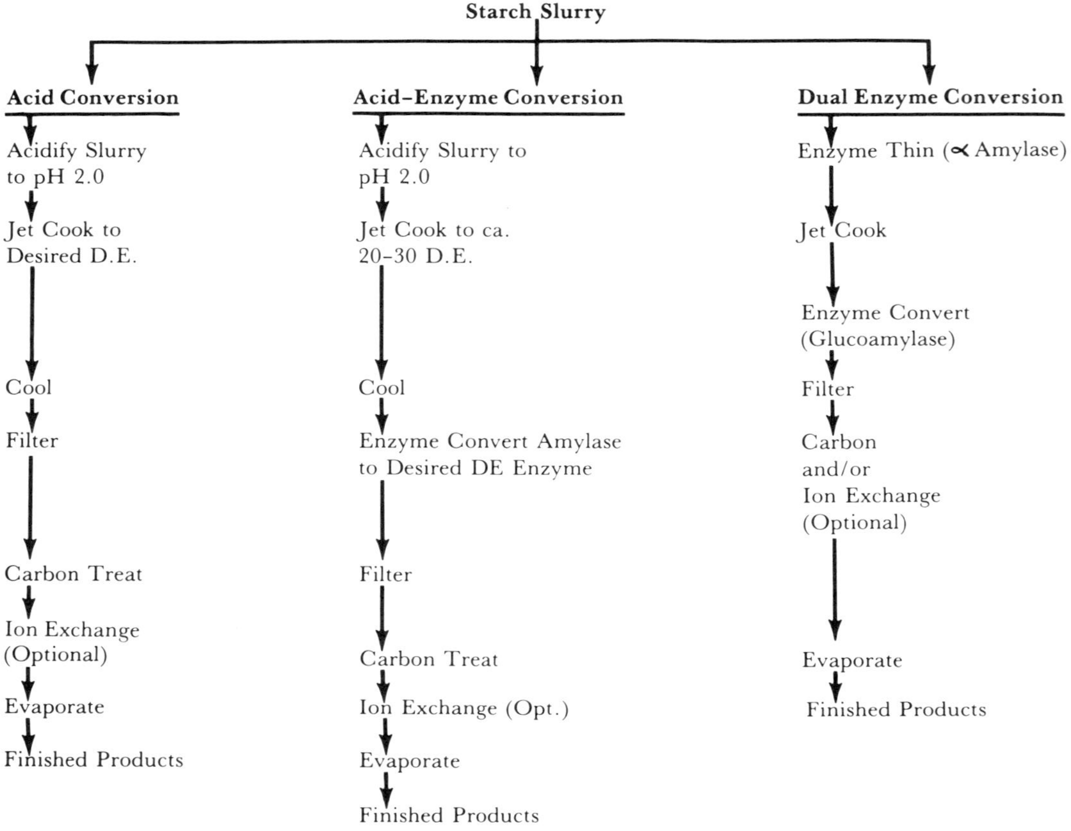

high molecular weight. β -amylase is very specific in that it hydrolyzes the starch chain at every other unit bond, producing large quantities of the disaccharide maltose. Glucoamylase yields large amounts of dextrose and therefore is important in the manufacture of high dextrose containing syrups. By using combinations of an enzyme (or enzymes) with varying acid hydrolysis treatments, a broad range of syrup products can be produced. Each one will possess chemical and physical characteristics which are the result of the effects of component saccharides taken as a whole.

Table I summarizes the most commonly produced regular corn syrups and includes the type of conversion process used to manufacture the product as well as a typical saccharide distribution for each and its D. E. value. The effects of these syrups on finished food product attributes will be discussed later in the chapter.

The high dextrose containing corn syrups (> 95% dextrose) are the substrate for the manufacture of the high fructose corn syrup products, and also for the dextrose crystallization process used to make crystalline dextrose monohydrate. Figure III is a process flow diagram for the production of high fructose corn syrups. The diagram shows a typical method for the production and subsequent refining of each generation of high fructose corn syrup. The isomerization step is the following enzymatic reaction: (see Figure IV).

FIGURE IV. ISOMERIZATION OF DEXTROSE TO FRUCTOSE

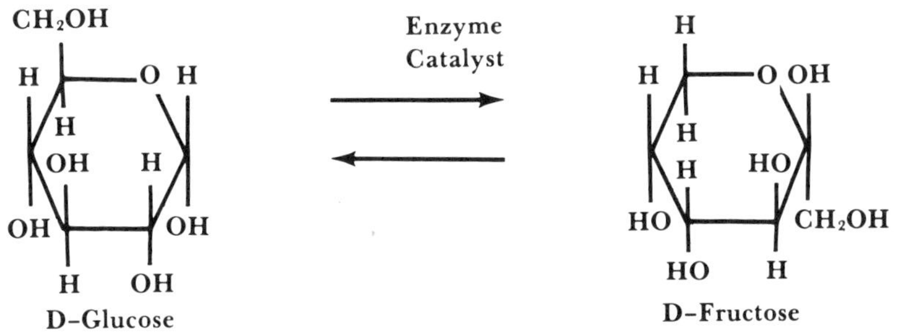

TABLE I

REPRESENTATIVE CARBOHYDRATE COMPOSITION AND CHEMICAL AND PHYSICAL DATA OF CONVENTIONAL CORN SYRUPS

Degree of Conversion	Very Low	Low	Regular	Intermediate	High
Type of Conversion	Acid-Enzyme	Acid	Acid	Acid	Acid-Enzyme
Dextrose Equivalent (D.E.)	26	35	43	54	64
Fermentable Extract (F.E.), %	23	32	42	54	76
Dextrose (Monosaccharides), %	5	14	20	30	39
Maltose (Disaccharides), %	14	12	14	18	33
Maltotriose (Trisaccharides), %	14	11	12	13	12
Higher Saccharides, %	67	63	54	39	16
Baume at 100°F., degrees	42	43	43	43	43
Total Solids, %	77.5	79.9	80.3	81.0	81.8
Moisture, %	22.5	20.1	19.7	19.0	18.2
pH	5	5	5	5	5
Acidity as HC1, %	0.015	0.015	0.015	0.015	0.015
Boiling Point, °F.	222	226	227	229	233
Weight, lbs./gal. at 100°F.	11.70	11.81	11.81	11.81	11.81

FIGURE III. ISOMERIZATION AND FRACTIONATION PROCESS FOR PRODUCTION OF HIGH FRUCTOSE CORN SYRUPS

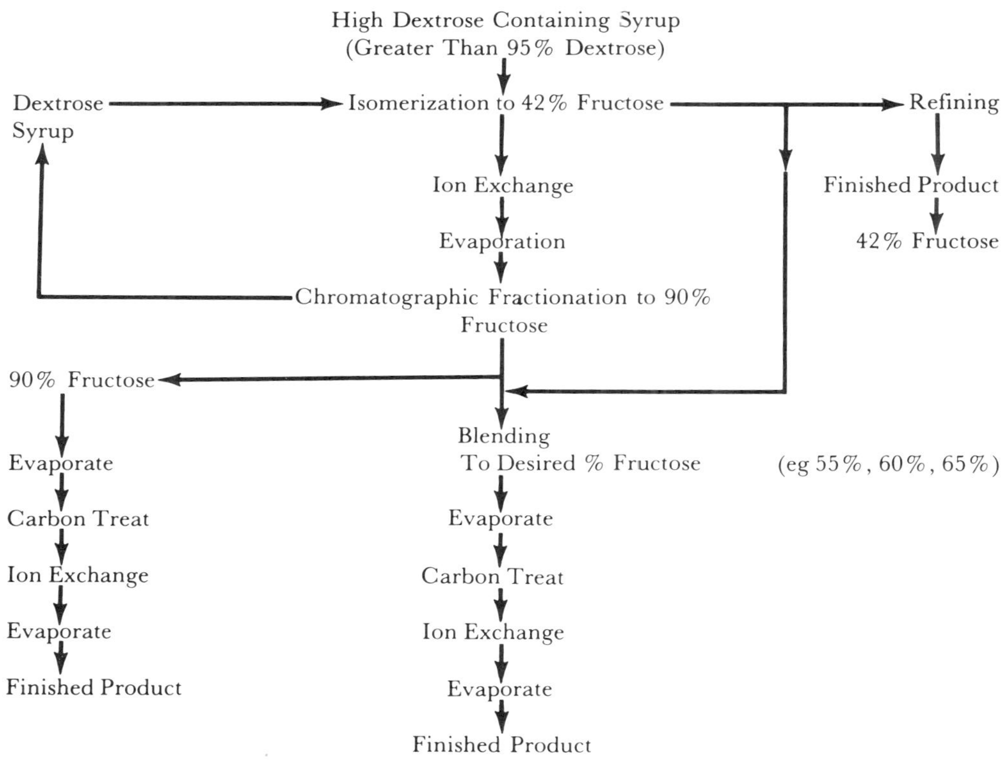

Being a reversible reaction, theoretically it would proceed to an endpoint balance of approximately 51.5:48.5 fructose to dextrose ratio. In commercial practice it is much more practical to stop the reaction when the fructose level meets meets 42%. Hence, this is the fructose composition for the first generation fructose syrups. By further separating off the fructose through a procedure based in theory to a chromatographic fractionation process, products with greater than 90-95% fructose can be achieved. These then are either sold as is or blended with syrups of lower fructose content to achieve a finished fructose content to meet the intended purposes. 55% fructose syrups (41% dextrose) have become most popular because of their similarity in functionality to inverted sucrose syrups. These syrups are commonly referred to as "second generation" high fructose syrups. The 90% fructose products are similarly referred to as the third generation of fructose syrups. The carbohydrate composition and some typical data on these syrups are shown in Table II.

III. FUNCTIONAL PROPERTIES

A. Sweetness

It has already been described that the D.E. of a given corn syrup is determined by the saccharide distribution within the syrup. The D.E. is only one of many factors which are the resultant effect of either individual saccharides or of the saccharides taken as a whole. Properties such as viscosity, osmotic pressure, fermentability, freezing point depression as well as others are all determined by the collective effects of the component saccharides in a syrup. The most common use for a corn syrup is as a sweetener, and this property too, is a product of the sweetness of the sacchrides that comprise the syrup.

The reasons for wanting a sweet character in a product are quite numerous as in today's market it is socially and psychologically acceptable and even advantageous to make products sweet. Each simple sugar or sweetener has an inherent level of sweetness which can be perceived differently depending on many factors in the environment in which it is employed. The sweetness of several common sugars has been rated organoleptically and compared to the sweetness of sucrose. This rating system is called

TABLE II

REPRESENTATIVE CARBOHYDRATE COMPOSITION AND CHEMICAL AND PHYSICAL DATA OF HIGH FRUCTOSE CORN SYRUPS

Corn Syrup (common designation)	HFCS 42	HFCS 55	HFCS 90
Carbohydrate Composition (% of saccharides)			
Fructose	42	55	90
Dextrose	52	41	8
Maltose	4	3	1
Higher Molecular Weight	2	1	1
	100%	100%	100%
Total Solids, %	71	77	80
Moisture, %	29	23	20
pH	4.0–4.4	3.5–4.0	3.5
Viscosity, Poises at 80°F.	1.60	6.70	11.00
Weight, lbs./gal. @80°F.			
lbs. as is	11.23	11.55	11.74
lbs. solids	7.98	8.83	9.39

"relative sweetness," and generally involves using a scale with sucrose as the standard having a relative sweetness of 100. The relative sweetness values for several common sugars are given in Table III.

TABLE III
The Relative Sweetness of Common Sugars[1]

Sugar	"Dry" or Crystalline State	In Sol'n. 10% water (w/w)
Fructose	180	100-175
Sucrose	100	100
Dextrose (α-D glucose)	76	65-70
Maltose	32	16-32
Galactose	32	16-32
Lactose	16	11-32

[1] ranges of values reported in the literature

As shown in Table III, the relative sweetness value for most sugars change when evaluated in aqueous solution. This is due in part to the mutarotation of the sugars from one form to another. β-d-fructopyranose, the crystalline form of fructose, is the sweetest of all sugars. In solution, however, the sugar mutarotates until an equilibrium is established between three other forms: α-d-fructopyranose, β-d-fructofuranose and α-d-fructofuranose. The less sweet fructofuranose forms account for the drastic apparent reduction in sweetness. Any factors which affect the rate of this reaction (the mutarotation) will also then affect the perceived sweetness. Fructose tastes sweeter in cold products than in hot products because more of the fructose exists in the β-d-fructopyranose form. The reverse phenomenon can be observed by putting crystalline fructose into a hot product such as coffee. The initial sweetness will appear quite high, but the sweetness will then decrease as the mutarotation takes place. Reports indicate that at 5° C, D-fructose solutions have a relative sweetness of approximately 144, while at 40° C and 60° C, relative sweetness decreases to 100 and 79 respectively.

Other factors which affect the perceived sweetness of many sugars include: 1. the solids concentration of the finished food product and 2. other ingredients which are present. High fructose corn syrup, for example, appear sweeter in foods that have a solids content greater than 20%. High fructose corn syrups also appear sweeter in foods that contain organic acids such as citric or malic rather than mineral acids such as phosphoric. All corn derived sweeteners have some sweetness value and are used for such. Table IV lists some relative sweetness values for various corn derived sweeteners. The high fructose corn sweeteners are comparable to sucrose (dsb) for sweetness in many applications. The presence of other sweeteners in the food may not only have an additive effect on the sweetness of an added sweetener, but synergystic effects are also believed to exist.

Table IV. Relative Sweetness of Corn Derived Sweeteners[a]

Syrup Designation	Relative Sweetness
HFCS-90*	106
HFCS-55*	99
HFCS-42*	91
63 DE	45
42 DE	30
High Maltose (44% maltose)	35
36 DE	25
Dextrose	67

*indicates % fructose
[a]Sucrose equal to 100

The sucrose inversion process which has been practiced widely for many years is a good example of how to make a single product into several more useful ones. When sucrose is inverted (or broken down into dextrose and fructose) several advantages are gained. These include a higher level of solubility (from 67% to 76.5% solids), an increased level of osmotic pressure and an increased level of sweetness. A similar situation exists in the case of corn syrup manufacture. The extent of conversion of the amylose and amylopectin molecules to lower molecular weight polysaccharides determines to a large degree the properties the resultant syrup will impart to a food product, and what advantages will be gained.

The most important functional properties (other than sweetness) associated with all corn syrups are: 1. viscosity; 2. osmotic pressure; 3. humectancy; 4. hygroscopicity; 5. browning; 6. fermentability; 7. freezing point depression and 8. flavor enhancement.

B. Viscosity

Corn syrups with a sizeable portion of higher molecular weight polysaccharides (greater than 30 dextrose units) have a higher viscosity than syrups made up almost entirely of monosaccharides. The long chains of molecules are less soluble and tend to form a network which, along with weak H+ bonds, holds moisture and thus increases the viscosity of both the syrup and the food product it is used in. Table V gives examples of the actual viscosities of a range of corn syrup products. By employing a certain syrup or blend of syrups many desired effects on the finished product can be achieved. The viscosity of the food can influence its "richness" organoleptic quality and can also directly add to the homogeneity and physical stability of the product. Corn derived sweeteners are used to achieve all of these qualities, and they serve as an inexpensive source of solids as well.

Table V. Comparative Viscosity Values of Corn Derived Sweeteners

Syrup Designation	Viscosity (Poises)* @100° F.
26 DE	220
35 DE	200
43 DE	125
42 DE High Maltose	125
54 DE	75
64 DE	55
HFCS 42% Fructose	0.75
HFCS 55% Fructose	2.70

*1 Poise = 100 Centipoises

C. Osmotic Pressure

The pressure exerted against a semi-permeable membrane when solutions of two different solids concentrations are placed on opposite sides of it is a practical definition of osmotic pressure. Water (or solvent) will then flow from the side with the lowest solids concentration to the side of high solids concentration. The amount of pressure which develops is a function of the concentration of the sugar or corn derived sweetener and its molecular makeup. The osmotic pressure is directly related to the number or molecules in solution. The more highly converted (higher D.E.) corn derived sweetener, including the high fructose corn sweeteners have the greatest effect on the osmotic pressure of a system. They are made up of a greater number of smaller (mono and di) saccharides than the lower D.E., lesser converted syrups. In fact, the pressures exerted by high fructose corn syrups, since they are approximately 95% monosaccharides, are about twice the pressure created by the disaccharide sucrose. This becomes important when considering equilibration rates for sweet pickle curing or maraschino cherry sweetening. Not only does the higher osmotic pressure of low molecular weight saccharides (such as fructose and dextrose) increase the rate of drawing water out of the vegetable or fruit, but the smaller molecular size also facilitates entry of the sugars back into the product. The higher osmotic pressure also lends the corn sweeteners themselves a significant amount of microbial stability, especially at higher solids levels. This stability does then carry through to high solids end product applications such as pancake and waffle syrups, mayonnaise and jams or jellies. This preservative action is related to the fact that micro-organisms have to compete with the water drawing osmotic pressure of the sugars to live. If this pressure is too great, the organisms have to compensate by slowing down their growth, not reproducing, and they may even die off. At similar solids concentration, the high fructose corn derived sweeteners are generally more microbially stable than liquid sucrose for this reason.

D. Humectancy

Corn syrups have long been recognized for their ability to retain moisture. In fact, the high D.E. syrups are very difficult, if not impossible to drum or spray dry. This feature does play an advantageous role in product where the retention of a small amount of moisture is desirable for softness or pliability. Intermediate moisture foods fall into this category. Low D.E. corn syrups are employed here to act as a "water sink," providing moisture for textural improvement, but also holding onto the water and preventing microorganism growth.

E. Hygroscopicity

Figure V illustrates the relationship between water adsorption-hygroscopicity and water desorption-humectancy. The hygroscopicity of a sweetener is also related to its saccharide composition. Syrups with a high number of monosaccharides tend to be very hygroscopic, which explains why candy manufacturers have to use lower D.E. syrups. They experience this as a problem when the acid in their formula and the heat in the process causes some of the formula sucrose to invert, yielding a given amount of highly hygroscopic fructose. The resultant candy is sticky and causes problems in packaging and on storage. High fructose corn sweeteners cannot be used in candies or coatings for this reason also. The use of low D.E. syrups to sweeten candies reduces the possibility of this problem occurring. Less hygroscopic lower D. E. syrup are ideal in this application, and are commonly used to control stickiness caused by the sucrose inversion. Graining, or the crystallization of sucrose in a high solids product can also be controlled by the addition of regular corn syrups.

FIGURE V. ADSORPTION-DESORPTION ISOTHERM

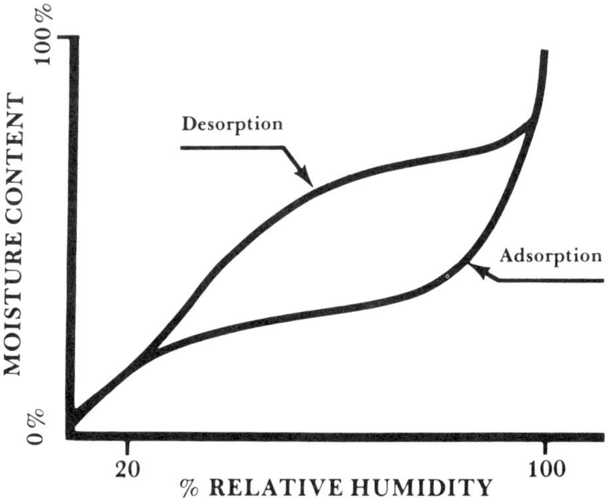

F. Browning

The development of color is an important function of corn derived sweeteners in a variety of food products. Bakers depend on the reaction between reducing sugars and protein (the non-enzymatic or Maillard browning reaction) to give their products a desirable crust color. Sucrose is a non-reducing sugar and, therefore, must be broken down into its component dextrose and fructose molecules before the Maillard reaction can take place. Higher D.E. corn derived sweeteners, as do the high fructose corn sweeteners, have a ready supply of reducing sugars in dextrose or dextrose and fructose respectively.

The reaction is, however, not always a desirable one. The usage of high D.E. syrups and the high fructose products is somewhat limited in several other applications where there is a significant quantity of protein available for the reaction, but color development is not desirable. Canned vanilla pudding is an example of this phenomenon. The reaction can be controlled to some extent by reducing the pH of the product to between 2.75 to 3.5 as is done in the production of various acidified dairy or soy based products. Also, since the rate of the reaction is temperature dependent (Q_{10} = 2-3), production and storage conditions play important roles in color development. (Shallenberger and Birch, 1975).

G. Fermentability

Fermentables, in our context, is defined as the percentage of carbohydrates in a sweetener which is fermented by baker's yeast, and is commonly expresed as % F.E. (fermentable extract). The F.E. can be approximated by adding the total mono and di-saccharide content. Polymers above DP_2 are usually not readily fermented. High fructose corn syrups have an F.E. of about 95. In yeast leavened breads then, HFCS functions very comparably with sucrose, total invert and/or dextrose. Similarly, lower D.E. corn syrups, which have a greater amount of higher molecular weigh polysaccharides are consequently lower in fermentability. High fructose corn syrups and dextrose are also used in ameliorating wines. The brewing industry has taken advantage of the total fermentability of dextrose and is utilizing it as a primary carbohydrate source for their popular light beer products. With dextrose, the residual carbohydrates normally left over after fermentation of the sugars in malted grain are not there, thereby reducing the calorie content of the final product.

H. Freezing Point Depression

Freezing point depression is related to the number of molecules in solution and becomes critical when formulating ice creams and other frozen desserts. The greater the number of molecules in solution, the more its freezing point will be depressed. Ice cream manufacturers have long been aware that the smaller the average molecular weight of their sweetener, the lower the temperature at which the mix freezes. For this reason, they effectively increase the average molecular weight by using a blend of a low molecular weight sweetener such as high fructose corn syrups (necessary for sweeetness) and a low D.E., high average molecular weight corn syrup.

The effects on the freezing point of four sweeteners are shown graphically in Figure VI. From this, it can be seen how the blending of a low D.E. corn syrup and a monosaccharide corn sweetener (dextrose or fructose) will balance out the freezing point of the product closer to that of a pure sucrose system. Use of an all monosaccharide sweetener, at a high enough concentration will actually depress the freezing point to such a degree that the product never becomes "frozen," in the classical sense. That is, so much water remains bound to the sweetener that there is so little free water available to form crystals and create a rigid structure. Less water is actually frozen at any given temperature. In ice creams, where ice crystal structure is important for body and texture, oftentimes an adjustment in stabilizers is necessary when large percentages of high fructose corn sweeteners are employed to tie up and physically immobilize the bound, but unfrozen water. This has the effect of balancing out for the extra amount of unfrozen water.

I. Flavor Masking - Enhancement

It is generally believed in the industry that lower D.E. corn syrups can be used to mask, or reduce the intensity of many flavors. Alternatively, the largely monosaccharide high fructose corn sweeteners and dextrose have been shown

FIGURE VI. FREEZING-POINT RELATIONS OF VARIOUS SWEETENING SOLUTIONS

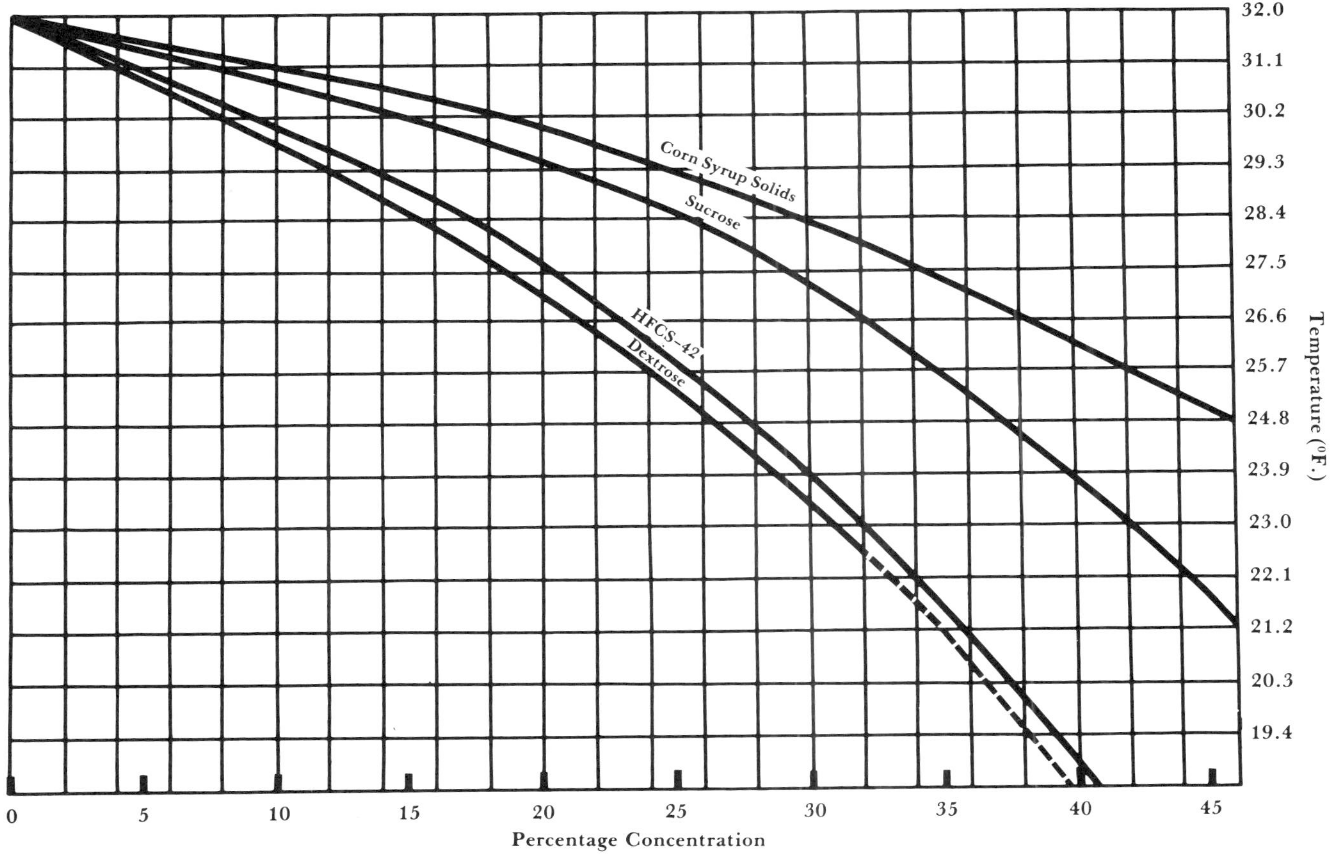

to enhance certain flavors. As examples of these, the low D.E. syrups are often used in conjunction with other sweeteners to tone down the effects of titratable acidity in salad dressings and/or mayonnaise. One theory suggests that possibly the large average size of the dextrose polymers physically bars the molecules of the acid from their taste receptor sites on the tongue.

The flavor enhancement quality of high fructose corn sweeteners is also thought to be closely related to molecular size. Being small molecules, the fructose and dextrose could possibly reach the taste receptor sites quicker than larger molecular sweeteners (such as sucrose). The resultant effect would be that the initial sweetness response is quicker, the peak sweetness is reached sooner, and the sweetness continues to dissipate earlier than the sweetness sensation of the disaccharide sucrose. The lingering sweetness of the sucrose is then thought to mask the appearance of certain later appearing taste sensations. Jelly makers have commented that their products taste "fruitier" when high fructose sweeteners are used to replace sucrose. Spices and flavor oils also appear more intense in systems sweetened with high fructose corn sweeteners, and, somewhat less intense in systems employing the low D.E. corn syrups.

J. General Properties in Summary

We have covered in a fair amount of detail the relationship between the component saccharides in a corn sweetener and the functional properties that result. Figure VII illustrates once again these properties and depicts the magnitude, from most to least (or vice versa) for each class of corn sweetener. From this figure, one can choose a particular sweetener which would best serve his needs for a specific application, or he can seek to obtain the optimum of several properties by blending two or more sweeteners. Although they are not included in Figure VII, dextrose and the high fructose corn sweeteners do exhibit the same properties. Due to the low molecular weight of their component saccharides, they will usually fall in line just to the right of the area under Highest Conversion Corn Syrup. In other words, dextrose and/or high fructose corn syrups have even less effect as a bodying agent than the higher conversion syrups, and more of a tendency to contribute to the browning reaction, as examples.

FIGURE VII. PROPERTIES AND FUNCTIONAL USES OF CORN SYRUPS

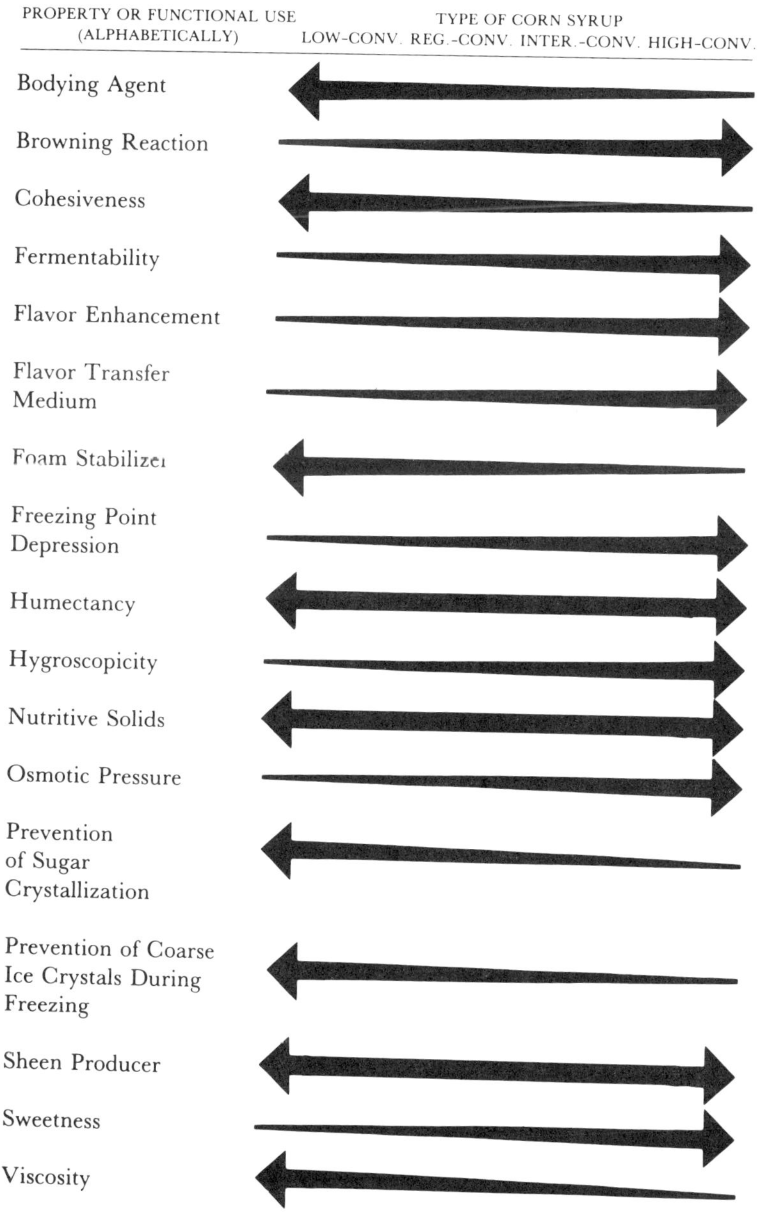

There are a number of corn derived sweeteners that have been developed especially for a very specific application. Most corn wet millers produce a variety of these specialty type products such as syrups high in dextrose and low in higher saccharides for brewers; syrups low in SO_2 for canners, and the high maltose syrups. The high maltose corn syrups were developed initially for use in candy and confectionery products to aid in preventing a high percentage of dextrose from forming and crystallizing out. They are used in jams and jellies for the same reason. The sucrose in these food products has a tendency to invert in the process and during storage. If the dextrose which is thereby formed is added to the dextrose content of the regular corn syrup also used in the food the total dextrose percentage may be greater than its limit of solubility for under the given conditions and crystallization will occur. The high maltose corn syrups, with only 7-10% dextrose, reduce the possibility of this occurring.

High maltose syrups are also excellent companion syrups for the high fructose corn sweeteners. Again, dextrose crystallization in jams, candies, and table syrups could be a problem when extensive quantities of the 42% high fructose product is used. To lower the overall percentage of dextrose, high maltose syrups are blended with the high fructose products. In this way a stable, sweet and controllable product can be obtained. Table VI shows characteristic information for two commonly produced high maltose syrups.

TABLE VI

CHARACTERISTIC DATA FOR HIGH MALTOSE CORN SYRUPS

Type	DE	% Dextrose	% Maltose	% Tri-Saccarides	% Highers
Regular Conversion	44	7	36	25	32
Intermediate Conversion	50	8	44	23	25

IV. APPLICATIONS SUMMARY

The applications of corn derived sweeteners evolve from the list of the functional properties previously discussed. The consideration of such measurable qualities as viscosity and reducing power is translated into more subjective organoleptic qualities such as mouthfeel, richness, and appearance. Taking all of the aforementioned properties into account, corn sweeteners are used to affect the organoleptic qualities of a food product in the following ways:

Organoleptic Quality	Mechanism(s)
Taste	Add sweetness Mask off-flavors Enhance desirable flavors Toner (balances flavors)
Aroma	Impart no inherent aroma Control odor release
Appearance	Stabilize particulates Impart surface sheen
Texture	Control body, viscosity Improve mouthfeel Add satiety value (richness) Add solids Retain moisture Prevent crystallization Impart pliability Improve spreadability
General	Extend shelf life and product quality Provide nutritive caloric value (approx. 4 Cal/gm dsb)

In addition, corn derived sweeteners are economical; produced from a domestic supply of a basic and plentiful raw ingredient. Most types are readily available throughout the United States, and either can be shipped direct from manufacturers in tank truck or rail car

TABLE VII

Corn Derived Sweetener	Applications
26 DE (liquid or dried)	Spray drying; coffee whiteners
35-36 DE (liquid or dried)	Coffee whiteners; whipped toppings; chewing gum; as a co-sweetener with high fructose corn sweetener in frozen dairy desserts; cake mixes; chewing gum
42 DE (liquid or dried)	Confections; fountain products; ice cream; preserves; table syrups; bakery fillings, bakery products; salad dressings; catsup
54 DE	Confections; marshmallows; table syrups; ice cream
63-68 DE	Intermediate and high moisture foods; ice cream; jams & jellies; salad dressings; fruit drinks; pickles; catsup; canned fruit
High Maltose	Canned fruit; jams and jellies, table syrups, toppings, confections
High Fructose	Carbonated beverages; ice cream; cured meats; chocolate milk; bakery products; wine; table syrups; jams & jellies; fruit drinks; pickles; canned fruit
Dextrose	Dry mixes; icings & donut sugars; tableting; baking; brewing

quantities, or obtained in smaller quantities through brokers and distributors.

In the final table, Table VII, a summary is given of the types of corn sweeteners that are available and some of their most common applications. These applications illustrate the broad range of products corn sweeteners are used in, but the list is not limited to them. Many times the end product characteristics quite obviously reflect the functional properties of the corn sweetener utilized. In any event, corn derived sweeteners are definitely becoming more of a standard, if not an absolutely necessary, ingredient in the foods we eat.

REFERENCES

Shallenberger, R. S. and Birch, G. G. (1975). Sugar Chemistry. Avi Publishing Co., Inc. Westport, Conn. p. 187.

Sugar and Sweetener Report, USDA. Vol. 6, No. 1, February, 1981. p. 40.

DAIRY BASED INGREDIENTS AND THEIR USE AS ALTERNATIVES TO TRADITIONAL SWEETENERS IN FORMULATED FOODS

Alan G. Hugunin

Foremost-McKesson Research Center
Foremost-McKesson, Inc.
Dublin, California

I. INTRODUCTION

Dairy based ingredients are used in formulated foods because of their nutritional, functional and economic benefits. Milk is a complex mixture of vitamins, minerals and unique proteins, fats and milk sugar. Although all dairy based ingredients are derived from milk, they differ significantly in composition, functional properties and cost. As a result, some food formulators are finding new applications for dairy based ingredients. Substituting dairy based ingredients for traditional sweeteners is one example of the new applications.

II. TYPES OF DAIRY BASED INGREDIENTS

The derivation of the most common dairy based ingredients is illustrated in Figure 1. Typical composition of many dairy based ingredients are listed in Table I.

ISBN 0-12-169080-6

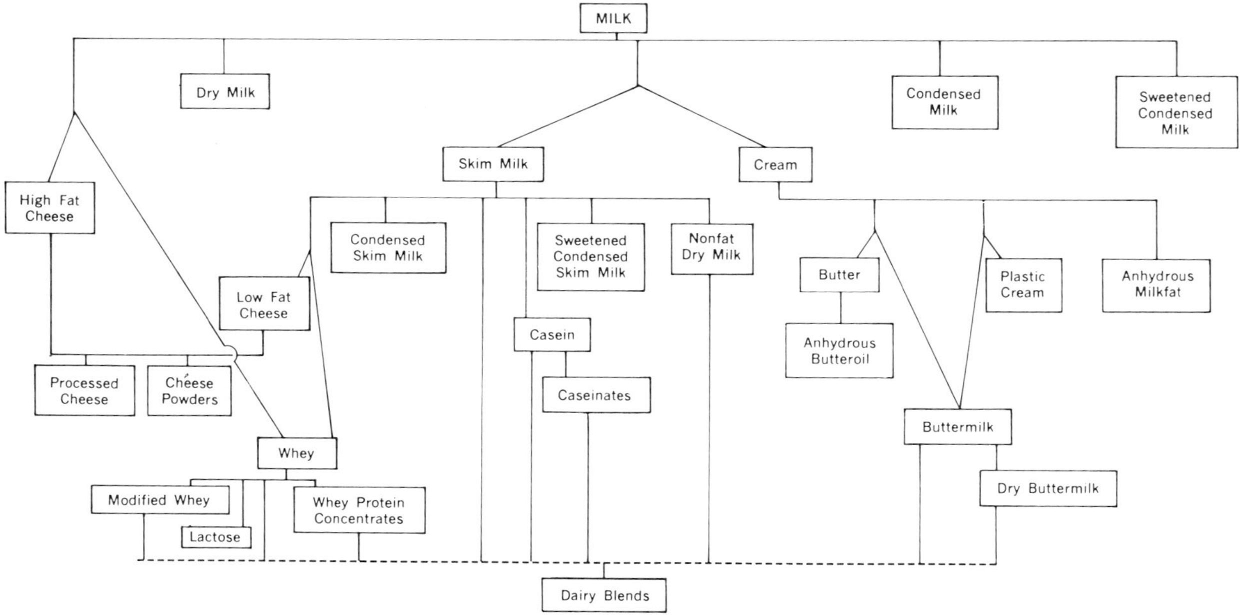

FIGURE 1: Dairy based ingredients from milk.

TABLE I. Typical Composition of Dairy Based Ingredients[a]

Ingredients	Moisture %	Protein %	Fat %	Lactose %	Ash %
Fluid whole milk	87.4	3.5	3.5	4.9	0.7
Fluid skim milk	90.5	3.6	0.1	5.1	0.7
Dry whole milk	2.0	26.4	27.5	38.2	5.9
Nonfat dry milk	3.0	35.9	0.8	52.3	8.0
Dry buttermilk	2.8	34.3	5.3	50.0	7.6
Cream, heavy whipping	57.3	2.2	36.8	3.2	0.5
Plastic cream	18.2	0.7	80.0	1.0	0.1
Butter	16.5	0.6	80.5	0.4	2.5
Anhydrous butteroil	0.2	0.3	99.5	0.0	0.0
Cheddar cheese	37.0	22.0	32.0	0-2.1	3.7
Cottage cheese curd	79.0	16.9	0.4	2.7	0.8
Casein	7	88	1	-	4
Caseinate	3-5	90-94	0.7-1.0	-	6-7
Dry whey	4.5	12.9	1.1	73.5	8.0
Reduced minerals whey	4.5	10-24	1-4	$\leq$85	1-7
Reduced lactose whey	4.5	16-24	1-4	$\leq$60	11-27
Whey protein concentrates	5.0	25-80	1-10	$\leq$60	2-15
Lactose	5.0	0.1	0.1	94.7	0.2

[a]From references 1-4

The use of centrifugal separators to produce high fat creams and nonfat milk products, churns to produce butter and buttermilk and acidification (microbiological or chemical) to produce whey and cheeses or casein are the traditional methods used to fractionate milk components. Almost everyone is aware of the identity of the resulting dairy products.

Modern technologies are being utilized to produce a new family of dairy based ingredients derived from dairy whey. Electrodialyses and ion exchange processes are used to produce reduced minerals whey. These whey products have lower salt or ash contents. Concentration, crystallization and crystal separation are used to separate lactose, the milk sugar. The co-product of lactose production is reduced lactose whey. Ultrafiltration, gel filtration or a combination electrodialysis and lactose separation processes are used to produce whey protein concentrates which contain not less than 25 percent protein. Effective September 4, 1981, these ingredients gained additional importance to food formulators. Whey, acid whey, reduced minerals whey, reduced lactose whey and whey protein concentrate were classified as GRAS (Generally Recognized As Safe) when used in accordance with "Good Manufacturing Practices" by the U.S. Food and Drug Administration (1).

III. NUTRITIONAL VALUE OF DAIRY BASED INGREDIENTS

Dairy based ingredients are recognized as wholesome products. Milk proteins have a balanced amino acid profile. Casein is the bench mark against which the protein efficiency ratio (PER) of other proteins are compared. Whey proteins which contain higher concentrations of the amino acids lysine and tryptophane have a higher PER value than casein. Most important, the amino acid profiles of these dairy proteins are complementary to the amino acid profiles of cereal and legume proteins. Thus, the nutritional value of many plant protein products can be significantly improved by blending with dairy proteins (5).

Calcium, phosphorous and magnesium are essential minerals in the human dietary. Many dairy based ingredients are rich in these minerals. Milk and milk products have been credited with providing approximately 72% of the calcium in the U.S. human dietary (6). The calcium to phosphorous ratio in cow's milk is 1.2:1. The optimum ratios of calcium to phosphorous in the diet have been identified as 1.5:1 for infants and 1.1:1 for adults (7).

High fat dairy based ingredients are excellent sources of vitamin A. Water soluble vitamins including thiamine, riboflavin, niacin, pyridoxin, pantothenic acid, biotin and folic acid are found in relatively higher concentrations in the nonfat dairy based ingredients. Approximately 36% of the riboflavin in the U.S. dietary is supplied by dairy products (6).

IV. FUNCTIONAL PROPERTIES OF MILK COMPONENTS

The functional properties of dairy ingredients vary significantly and depend upon the concentration of the individual milk components in the ingredient. It is frequently possible to identify the specific functional properties required in a formulated food and match that to the composition of various dairy based ingredients.

Caseins have high water binding capacity and are effective stabilizers for fat in water emulsions. They add strength to dough systems and coagulate to form a curd structure in acidic systems.

Whey proteins have low water binding capacity and remain soluble in acid, neutral and alkaline systems. In concentrated solutions, whey protein will gel at high temperatures. They have excellent whipping, foaming and emulsifying properties but are less effective than caseinates as stabilizers. Whey proteins generally produce a tenderizing effect in dough systems.

Milk fat is highly regarded because of its flavor contribution to formulated foods. Also, milk fats can improve the texture of other fats. For example, small quantities of milk fat can reduce the "waxy" mouthfeel obtained with certain vegetable fats. Phospholipids and lipoproteins, which make up the fat globule membrane in milk, are very effective natural emulsifiers.

Lactose is a dissacharide and is the predominant individual component in milk and many nonfat dairy based ingredients. Although it is a highly functional ingredient in many formulated foods, its attributes are probably the least well recognized of the dairy components. Its functional properties are unique and should become apparent as the applications for dairy based ingredients are discussed.

V. APPLICATIONS FOR DAIRY BASED INGREDIENTS AS ALTERNATIVES TO TRADITIONAL SWEETENERS

Traditional sweeteners used for formulating foods include sucrose, glucose, fructose and various types of corn syrups. These ingredients provide a number of functional benefits in formulated foods. Some of the most important functional roles are listed in Table II.

TABLE II. Functional Roles of Sweeteners

1. Add sweetness
2. Add texture and mouthfeel
3. Act as bulking or dispersing agents
4. Enhance and stabilize flavors
5. Preservation
6. Retain moisture
7. Develop color and flavor
8. Act as drying and agglomerating aids
9. Act as tableting excipients
10. Substrates for fermentations

Certain low fat milk products, such as nonfat milk, various whey products and lactose, can produce similar or desirably different functional properties when utilized in combination with or as an alternative to the traditional sweeteners. This is exemplified by considering dairy based ingredients in each functional role.

A. Sweetness of Dairy Based Ingredients

Sweeteners are typically considered in terms of the sweetness they impart. Compared to sucrose, lactose is relatively low in sweetness as is shown in Table III. However, the relative sweetness increases as the concentration of lactose increases. It can be further increased to approximately 70% of sucrose by hydrolyzing the disaccharide and producing glucose and galactose. Although hydrolysis is reportedly becoming a common commercial process in Europe, it is still in the development phase in the United States. In most current applications for dairy based ingredients, the relatively low sweetness can be considered and attribute. Sweetness does not become the limiting factor. High concentrations of dairy based ingredients can be used to maximize the other functional benefits.

TABLE III. Relative Sweetness of Sugars as Percent Concentrations to Give Equivalent Sweetness[a]

Sucrose	Glucose	Fructose	Lactose
1.0	1.8	0.8	3.5
5.0	7.2-8.3	4.2-4.5	13.1-15.7
10.0	12.7-14.6	8.6-8.7	20.7-25.9
15.0	17.2-20.0	12.8-13.0	27.8-34.6
20.0	21.8	16.7	33.3

[a]From reference 8

B. Effect of Dairy Based Ingredients on Texture and Mouthfeel

Increasing the solids concentration of a formula generally improves the texture and mouthfeel. Higher solids contents will produce thicker sauces and gravies, firmer dips and spreads and richer tasting puddings, frozen desserts, icings, toppings and beverages. The concentration of sucrose or glucose in these products is limited by the sweetness which they impart. Starches or gums can produce a perceived effect of higher solids, but the same texture is not always obtained. Dairy based ingredients used alone or in combination with other ingredients will frequently provide the desired effect. Nonfat milk or whey and caseinate blends will bind sufficient water to provide firmer textures. Whey products and lactose function similar to sugar. The improvement in texture is a direct result of the increased solids and not a result of the water binding. Electrodialized whey or lactose can be used at reasonably high concentrations in food products without producing an excessively sweet taste.

Canned icings are a prime example of products which many consumers find to be excessively sweet. However, it is necessary to incorporate a relatively high concentration of solids in these oil based products to produce a desirable flavor and texture. Reduced minerals whey can provide 10-20% of these solids, add milky flavor, enhance the emulsification due to the functionality of the whey proteins and reduce the sweetness.

Sweeteners can also be used to improve the texture of other types of foods. Sucrose will help to produce firm pickles, but the potential improvements are limited by the limitations on the concentration of sucrose. Jelen and Breene reported that lactose, because of its low sweetness, could be used at more effective concentrations and thereby improve the brittleness, hardness and elasticity of pickles (9). This may have applications in other vegetables.

C. Dairy Based Ingredients as Bulking and Dispersing Agents

Many new foods are being introduced as dry, shelf stable, convenience products. Sauce mixes, gravy mixes, seasoning blends, salad dressings, soups, beverage products, seasoned noodle lunches, and even packets of artificial sweeteners all require bulking and dispersing agents to carry the relatively small quantities of seasonings, colors and flavors. It is very desirable to have nonhygroscopic carriers which will not cake. The carrier should facilitate mixing and ideally it should dissolve in water. Traditional sweeteners can be utilized, but they may produce an atypical sweet taste in many products. Whey, reduced minerals whey and lactose generally meet all of the required criterial and frequently provide a flavor benefit. It is now possible to obtain agglomerated forms of these products which have a lower bulk density and disperse more readily.

D. Stabilizing and Enhancing Flavors with Dairy Based Ingredients

It is common knowledge that sugars, at flavor threshold levels, will enhance many flavors. Likewise, lactose has been found to be a very effective flavor enhancer and has been compared to monosodium glutamate (MSG) in this function (10). Most likely the mechanism is the same as that of other sugars; however, higher concentrations of lactose can be used without sweetness becoming the predominate flavor note.

Acid to sweetness ratios are often difficult to control in foods. Frequently the result is a separate and distinct sweet and sour flavor impact. Better blended flavors are possible with lactose because it does not produce the distinctly sweet taste sensation. Unpublished studies carried out in our laboratory found that tomato products could be effectively mellowed with a simultaneous enhancement of tomato flavor by adding 2-3% lactose.

The ability of lactose to absorb flavors has been well documented. Nickerson and Dolby reported that lactose anhydrides, when exposed to odors in chocolate and instant coffee plants, showed a greater capacity to absorb flavors than other sugars. (11). At least two Swiss patents were issued in the early 1950's which were based on the ability of lactose to preserve the aroma of roasted coffee in extracts (12). This property can be used advantageously in preparation of dried, flavored ingredients and foods.

E. Dairy Based Ingredients as Preservatives

Sweeteners are frequently used as preservatives in products such as jams and jellies. In many of these products, the fruit flavors are masked by the intense sweetness. Although lactose would appear to provide a solution, its use is restricted by its relatively low solubility. As is shown in Figure 2, lactose is less soluble than sucrose or glucose. Without sufficient water, it will crystallize and produce a gritty texture. Reger proposed that adding finely ground lactose would produce a myriad of impalpable crystals and produce a texturally acceptable product with improved flavor (13). Preserves prepared with 15-20% of the sweetener being lactose reportedly will also retain better color (14).

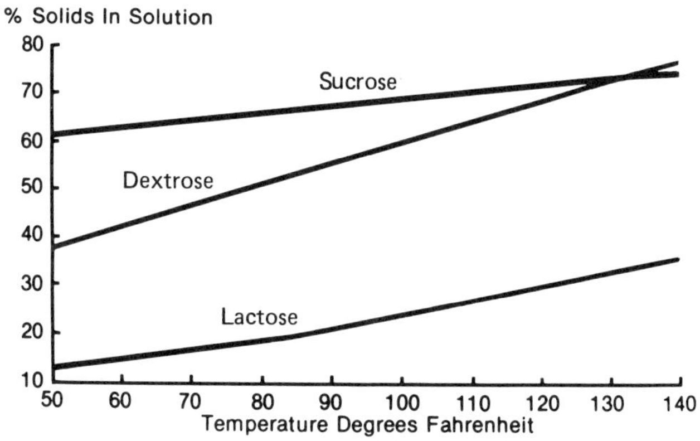

FIGURE 2. Solubility of lactose, sucrose and dextrose.

Nickerson studied the effects of lactose on the crystal habit of sucrose. He found that in combinations, the crystal habit of both sugars were changed. Whereas sucrose crystals normally tend to grow and interlock, thereby producing hard, course textures, a softer, smoother crystalline mass can be produced by mixing lactose and sucrose (8). This offers potential in certain candy and confection products which tend to harden with age. Whey, reduced minerals whey and lactose are already used by a number of confectioners producing caramels, fudges and cream center candies.

F. Effect of Dairy Based Ingredients on Moisture Retention

Sweeteners are frequently utilized to reduce moisture loss. Patel and Nickerson reported the rate of water loss from solutions in ascending order was sucrose, lactose, glucose and pure water (15). In mixtures of sucrose and lactose, glucose and lactose and glucose and sucrose the rate of moisture loss varied only slightly but did increase as the ratio of lactose or glucose increased. Interesting is their observation that after crystallization of lactose, the rate of water loss temporarily increased and then decreased.

Holmes and Lopez observed that breads and cakes formulated with lactose replacing a portion of the sucrose and shortening were initially softer and retained their softness over a longer period of time (16). As is shown in Figures 3 and 4, compressimeter readings over a five day period indicate shelf life can be more than doubled. They hypothesized that the unique water binding properties of crystalline lactose were responsible. In this particular application the water binding properties of lactose appear to be superior to sucrose.

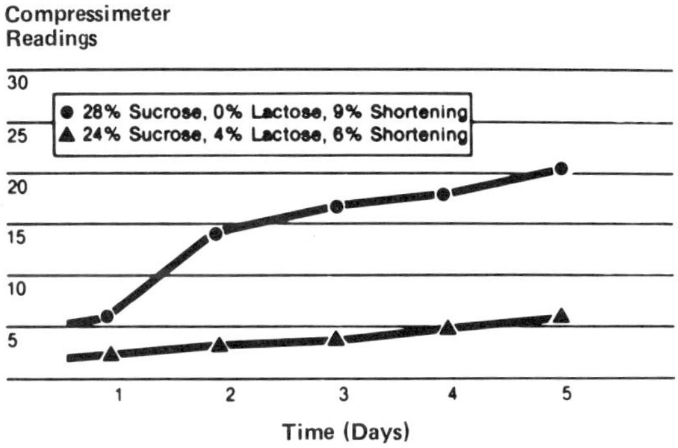

FIGURE 3. Bread compressimeter readings over five days as a measure of tenderness and shelf-life (16).

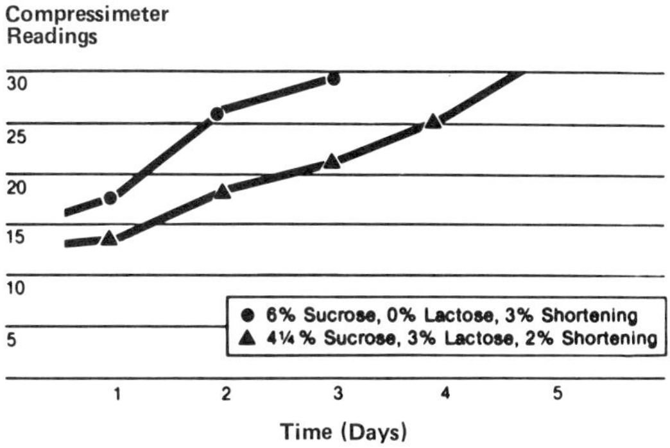

FIGURE 4. Cake compressimeter readings over five days as a measure of tenderness and shelf-life (16).

G. Color and Flavor Development by Dairy Based Ingredient

Sweeteners produce many of the flavors and colors associated with a number of food products through caramelization and nonenzymatic browning reactions. Lactose is a reducing sugar and will undergo both types of reactions. Whey and reduced minerals whey contain high concentrations of lactose plus the amine groups for the nonenzymatic browning reaction. The color and flavor of breads, cakes, caramels and toffees, custards and other similar foods can be enhanced through the use of these dairy based ingredients. A process as simple as rinsing potato slices in a whey solution prior to frying will enhance the golden color (8). With continually rising costs of energy, increasing the concentration of whey products in bakery formulations offers a means of obtaining the desired crust color with shorter or lower temperature bakes.

H. Dairy Based Ingredients as Drying and Agglomerating Aids

Dried products are gaining popularity because they have long shelf life and because they are more economical to ship and store. Frequently it is beneficial to increase the solids concentration of a solution to facilitate drying. Sweeteners can be used to increase the solids content. However, they are difficult to spray dry because of their low melting points. Compared to other common sweeteners, lactose has a relatively high melting point and can be spray dried with relative ease (Table IV). Combined with the other properties, such as low sweetness, the ability to absorb flavors and to serve as a bulking and dispersing agent, lactose and high lactose ingredients such as reduced minerals minerals whey are ideal drying aids.

Agglomerating is a process which is also increasing in popularity. It greatly facilitates the dispersion of dried products and will minimize stratification of ingredients in packages. Sweeteners can be used but lactose and whey products generally provide the optimum rewet and adhesive properties to produce non fragile yet readily dissolvable agglomerates.

TABLE IV. Melting Points of Sugars

Sugar	Temperature, °C
α-Lactose	202
β-Lactose	252
Galactose	167
Sucrose	160-186
Glucose	146
Levulose	102-104
Maltose	102-103

I. Dairy Based Ingredients as Tableting Excipients

Tablets are frequently associated with the pharmaceutical industry, but they have in recent years gained importance in foods. Flavored and colored tablets are being sold as candy. Numerous nutritional supplements ranging from fibers to proteins are being sold in tablet form. Ingredients which are to be used in relatively low concentrations are being tableted to provide portion control and eliminate the need for weighing. Lactose is one of the most common tableting excipients in the pharmaceutical industry. Spray dried lactose and certain dried whey products can be easily tableted with a direct compression tablet press and produce firm but easily dissolvable tablets.

J. Dairy Based Ingredients as Fermentation Substrates

Lactose is a fermentable sugar and can be converted to ethyl alcohol, lactic acid and single cell proteins. However, not all organisms have the ability to metabolize lactose. Bakers yeast for example cannot metabolize lactose, and one must include other fermentable sugars for proper leavening. A resulting advantage is that lactose remains intact as a reducing sugar for crust color development during baking. Other organisms can utilize lactose but at a slow rate. This can be an advantage because it restricts cell growth rate and increases the overall microbial production of desired chemicals. The less expensive dairy based ingredients such as whey are being evaluated as ingredients for beer production, wine production and ethanol for industrial uses (18,19).

VI. CONCLUSIONS

Dairy ingredients will continue to provide food formulators with the nutrition, flavor and texture attributes which homemakers obtain from milk. The new technologies which have developed in the dairy industry and the new lines of products which have evolved enable the food formulators to optimize cost and maintain quality originally obtained with milk. These new technologies have also reduced the cost and modified the functionality of some dairy based ingredients making them alternatives to other food ingredients such as sweeteners. Dairy based ingredients including whey, reduced minerals whey and lactose provide the food formulators with new options to creatively add nutrition, enhance or modify colors and flavors, modify textures or simply reduce costs.

REFERENCES

1. U.S. Dept. Health, Education and Welfare, Food and Drug Administration, Fedl. Register 46(172, Sept 4), 44434 (1981).
2. Hargrove, R., and Alford, J., _in_ "Fundamentals of Dairy Chemistry," 2nd. ed. (B. Webb, A. Johnson and J. Alford, ed.), p. 58. AVI Pub. Co., Westport, CN (1974).
3. Nutting, G., _in_ "By-products from Milk," 2nd. ed. (B. Webb and E. Whittier, ed.), p. 1. AVI Pub. Co., Westport, CN (1970).
4. Watt, B. and Merrill, A., "Composition of Foods; Agriculture Handbook No. 8." U.S. Government Printing Office, Washington, D.C. (1963).
5. Hugunin, A. and Lee, S., _Food Processing_ 36(11), 40 (1977).
6. Milk Industry Foundation, "Milk Facts" Milk Ind. Foundation, Washington, D.C. (1981).
7. Food and Nutrition Board, National Academy of Sciences, "Recommended Dietary Allowances," 8th. ed., National Research Council, Washington, D.C. (1974).

8. Nickerson, T., J. Dairy Sci. 59, 581 (1976).
9. Jelen, P. and Breene, W., J. Food Sci. 38, 99 (1973).
10. Keay, J., Mod. Dairy 50(3), 24 (1971).
11. Nickerson, T. and Dolby, R., J. Dairy Sci. 54, 1212 (1971).
12. Tweedie, L., Macbean, R. and Nickerson, T., Food Technol. Aust. 30(2), 57 (1958).
13. Reger, J., Cereal Sci. Today 3(10), 270 (1958).
14. Anon. Food Proc. 17(11), 22 (1956).
15. Patel, K. and Nickerson, T., J. Food Sci. 36, 495 (1971).
16. Holmes, D. and Lopez, J., Baker's Digest 51(1), 21 (1977).
17. Foremost-McKesson, Inc., "Lactose, Physical and Chemical Properties", Foremost-McKesson, Inc., San Francisco, (1979).
18. Jacobs, L., Food Prod. Devel. 15(3), 42 (1981).
19. Lang, F. and Lang, A., Milk Ind. 81(11), 30 (1979).

FABRICATION OF LOW-MOISTURE, HIGH-PROTEIN FOODS USING SOY ISOLATES AND HTST TWIN-SCREW EXTRUSION

Eugene Sander
Ron Bennett

Technical Service Department
Grain Processing Corporation
Muscatine, Iowa

Fred Austin

Pembertons
Food Processing Equipment Inc.
Memphis, Tennessee

I. INTRODUCTION

Fabrication of low moisture, high protein foods using soy isolates and HTST twin-screw extrusion was a project undertaken at Grain Processing Corporation to add new dimensions to soy isolate utility. This investigation demonstrated the first successful texturization of soy isolates using HTST extrusion technology.

CHEMISTRY OF FOODS AND BEVERAGES:
RECENT DEVELOPMENTS

ISBN 0-12-169080-6

The main objectives of this study were:

1. Optimize formulation and process control to:

   A. Establish good final product palatability.

   B. Generate product application flexibility such as development of easily-flavored functional extenders.

   C. Develop an ingredient with an acceptable price/value.

2. Replace textured soy flours and concentrates in new and existing products where a bland-flavored and functional extender is required to improve overall palatability.

3. Texturize 90% protein compositions utilizing HTST technology.

Four phases of this study will be presented in this paper including formulation of textured high protein compositions, application of HTST technology, texture profile analysis of resulting products, and development of half-product analogs.

II. FORMULATION

Figure 1 depicts the formulation of half-products with controlled compositions. Half-products are defined as those partially fabricated compositions which need further processing to put them in final palatable form. In addition to soy isolates, other protein choices can be readily used in formulating a functional, high protein textured product. Ingredient choice is not limited to just proteins; carbohydrate sources, such as rice flour, wheat flour, or fiber, can be included. Additives such as flavor, fat, maltodextrins, and other textured modifying agents such as sulfur can be included in the formulation. Building performance through ingredient choice and use of additives to enhance final product functionality and character are two essential goals in formulation.

FIGURE 1.

Formulation of Half Products with Controlled Compositions

Blends	With Additives
Protein/Protein	Flavor
Protein/Carbohydrate	Fat
Rice Flour	Maltodextrin
Wheat Flour	
Fiber	

Most of the work in this study was conducted using Grain Processing's line of soy protein isolates. PRO-FAM® choices included S-970, S-640, S-636, S-955, S-950, S-901, and S-792. One opportunity in this study was to relate isolate functionality to final textured half-product performance and functionality. To achieve this, isolates were selected on the basis of NSI (Nitrogen Solubility Index). By definition, NSI is determined by slurrying a six percent isolate solution followed by a 20-minute centrifugation; percent protein in the supernatant is then determined. Other PRO-FAM® choices included enzyme-modified (S-792) and iso-electrically precipitated (S-901) soy protein isolates. Table 1 presents the physical/chemical properties of PRO-FAM® soy protein isolates. Three PRO-FAM® products, S-970, S-640, and S-955 were selected on the basis of their decreasing NSI values. Protein, moisture and color values for the respective isolates are quite comparable; bulk density values decrease as the NSI value increases. Reported protein and moisture values were determined by standard methods of analysis. Bulk density was determined by filling a 100-cc graduated cylinder with material, weighing, jogging for three minutes on a Syntron jogger, and reading the final volume in cc's. Color was determined using a Hunter color meter and color difference values L,a, and b were measured; only L values (which reflect brightness or lightness) are reported in Table 1. Material leaving the HTST extruder was collected from a ribbon die and cut into 3.81 x 7.62 cm. strips; randomly sampled strips were kept for subsequent bulk density determinations; random samples of the ribbon were size reduced and passed

TABLE 1. Physical Chemical Properties of PRO-FAM® Soy Protein Isolates.

Variable	PRO-FAM S-955	PRO-FAM S-640	PRO-FAM S-970
N.S.I.	20	55	85
Protein (%) d.b.	91	91	93
Moisture	5.0	4.2	4.1
B.D.* Bulk Density	44	40	34
Color†	80.2	81.8	80.6

*Packed, (gm/100 cc)
†Hunter Colorimeter - L Value

through a 12-mesh screen. This size (<12 mesh) material was used to determine the (ground) bulk density values reported in Table 3. Bulk density values for the cut strips (chunks) were determined by placing a 10 to 25-gm. sample (made up of several cut strips) into a conventional bread loaf volume measuring device. The volume of the strips was determined by rape seed displacement.

Hunter color values were determined on randomly cut strips, size reduced and sized through a 12-mesh screen.

Rehydration characteristics of the textured half-products were determined by placing a 10-gm. sample of 2.54 x 2.54 cm. pieces into a beaker; excess water was maintained in the beaker for one hour. At the end of that hour, the hydrated pieces were drained to remove excess water and the hydrated pieces were weighed. The weight of the dry sample plus water was divided by weight of the original dry sample to determine the moisture/solids ratio.

Retorting characteristics of textured half-products were determined as follows; 15-gm. samples of 2.54 x 2.54 cm. pieces were placed in pint Mason jars; 300 mls. of water were added to the jar. Retorting was completed after 15 minutes at 2.02 atmospheres. Following cooling, the solids and liquid were separated and their respective weights determined. Hydrated solids were saved for texture profile analysis.

Texture profile analysis was determined using the Instron universal testing machine and the Kramer compression-shear cell; a typical two-bite compression-shear curve is shown in Figure 2. Two-bite curves were determined by placing 15 grams of 2.54 x 2.54 cm. pieces (randomly placed) in the Kramer cell. Measurement of three physical characteristics of the textured half-products was made;

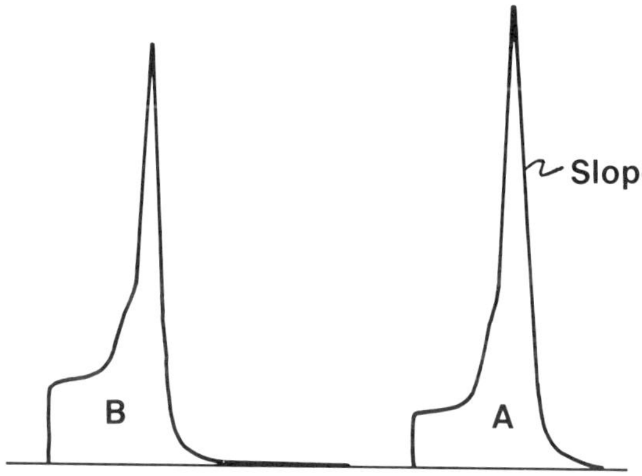

FIGURE 2. Compression-Shear Force, Two-Bite Texture Profile Analysis on Half-Product Analogs-Performed Using the Kramer Shear Cell.

cohesiveness is objectively defined by the ratio of the area under curve B to the area under curve A. Cohesiveness defines internal bonding strength. Hardness is determined by measuring the slope of the ascending leg on curve A; resulting force/distance values are reported in Kg./cm. Hardness is defined as the force required to cause deformation of a material. The third parameter, compression-shear force, is determined by the measured area under curve A; it objectively defines the energy (chewing) required to masticate foods using natural compression/shear/extrusion action.

III. HTST TWIN-SCREW TECHNOLOGY

The Creusot-Loire (C/L) extruder was used to prepare all half-product analogs described in this study. The basic unit consists of a separate control panel used to monitor all process parameters. Separate feeding of the dry solids and liquids to the barrel is key to good steady state operation; this eliminates the need for any prewetting of the feed material entering the extruder. Two rotating, self-wiping screws are contained within a barrel that is jacketed for cold water cooling at different points. Endothermic heat is provided by an electromagnetic induction heater. This system is 80% efficient and relies on barrel construction materials which allow for rapid heat transfer to the inside three mm. of hard steel barrel wall closest to the turning screws. The combination of electromagnetic induction heating, jacketed water cooling, and barrel construction materials permits a rapid system response to command signals for heat addition or removal. Successful half product modification using the twin-screw extruder is also determined in part by the choice

FIGURE 3.

C/L Extruder
Screw Profile/Action Zones

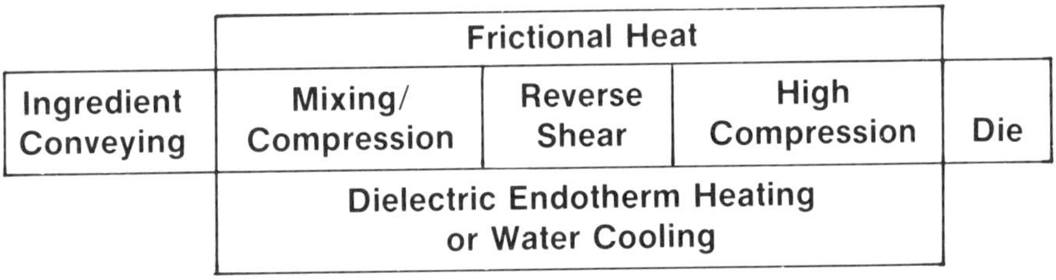

of screw profile used. Figure 3 illustrates typical C/L extruder screw profile/action zones. Through an ingredient conveying zone and on into a mixing compression zone, the product experiences the beginning of physical modification. Continuing on through a reverse shear and high compression zone area, the product moves on into a decompression zone where volatiles can be flashed off, or other ingredients added. Before reaching the die, generated frictional heat and endothermic heat further influence the physical changes the half-product analog undergoes within this environment. As the half-product leaves the die, it undergoes further physical and chemical changes as excess water flashes at the die face and the product experiences expansion texturization.

Figure 4 defines typical reverse-shear screw profile construction. As indicated, the numbers relate to the cross-sectional slot opening on the flights, number of slots in the screw, and pitch (angle degrees) of the flights. Proper selection and location of the reverse-shear section is a key to generation of successful changes in the physical/chemical characteristics of the finished product.

The advantages of an HTST twin-screw extruder include: 1) flexibility of operation, 2) responsive process parameter control, 3) low cost of operation, and 4) the elimination of

FIGURE 4.

C/L Extruder

Reverse Shear Definition

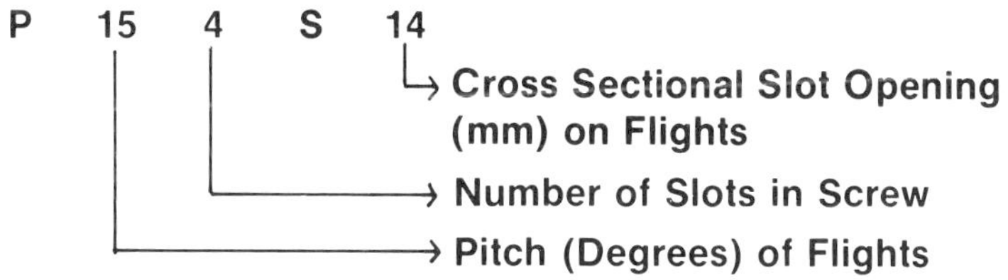

the need for post-extrusion drying of the finished product. As stated earlier, only HTST twin-screw extrusion technology has been successfully used to texturize 90% protein half-product compositions (100% soy protein isolates).

IV. PRODUCT DEVELOPMENT RESULTS

A summary of conditions used to establish NSI effects on PRO-FAM® half-product characteristics is presented in Table 2. Die opening, screw profile, screw speed, drive energy, barrel temperature, melt temperature, melt pressure, and feed rate of dry solids and water were measured. A summary of the values used is presented; the drive energy, melt temperature, and melt pressure vary as the NSI changes. Melt pressure and melt temperature increase as the NSI value decreases.

TABLE 2. HTST Extruder Conditions Used to Establish the Effect of NSI on PRO-FAM® Half-Product Characteristics.

Variable	PRO-FAM S-955	PRO-FAM S-640	PRO-FAM S-970
Die Opening (mm)		(1.5 × 22mm)	
Screw Profile		Std. with 4S14 Reverse	
Screw Speed (rpm)		200	
Drive Energy (amps)	19	21	18
Barrel Temperature (°C)		200	
Melt Temperature (°C)	184	176	182
Melt Pressure (kg/cm^2)	1350	1250	950
Feed Rate			
Dry Solids*	2.5	3.2	3.2
Water (l/hr)	4.4	4.4	4.4

*2.5 adjustment required by bulk density.

Changes in NSI and bulk density values for extruded PRO-FAMS® using constant HTST extruder conditions are shown by the data presented in Table 3. NSI values are reduced to values ranging from 3 to 11; compared to the higher starting NSI values, one can conclude that measured NSI value does not reflect either the physical or chemical changes that have taken place during the HTST extrusion process. Variations in bulk density on the ground (<12 mesh) and chunk (strip) half-product are also presented in Table 3. Bulk density values for ground samples increase after extrusion of the highest (S-970) and lowest (S-955) NSI value soy protein isolates.

The opposite is true when the intermediate NSI isolate (S-640) is extruded; this reversal cannot be explained at this time. To reflect the sensitivity and responsiveness of the HTST extrusion system to slight changes in formulation, a second trial on PRO-FAM® S-640 was conducted. A simple change in moisture/solids ratio from 0.81 to 0.92 caused a significant increase in both the ground and chunk (strip) bulk density values. For this reason, all controllable extruder conditions were held constant to fairly assess the effect of NSI value on extruder and resulting textured isolate performances. A trend upward in bulk density values was established by the textured strip material as the NSI value decreased (going from S-970 to S-955). One can conclude from this finding that NSI is a factor in selecting the starting

TABLE 3. Changes in NSI and Bulk Density Values for PRO-FAM® Half-Products Prepared Using Constant HTST Extruder Conditions.

PRO-FAM®		NSI	Bulk Density (gm/100 cc)	
			Ground[1]	Chunk[2]
S-970	As is	85	34	—
	Extruded	8	42	10
S-640	As is	55	40	—
	Extruded	11	24	11
S-955	As is	20	44	—
	Extruded	3	64	35
S-640	Extruded[3]	7	43	20

[1] Ground to ≮12 U.S. std. mesh screen—packed bulk density.

[2] 3.81 cm × 7.62 cm piece size—loose bulk density by rape - seed displacement.

[3] Moisture/solids ratio changed from 0.81 to 0.92.

soy protein isolate for subsequent conversion into a textured isolate half-product; desired physical/chemical characteristics and final half-product performances are influenced by starting NSI values.

Influence of the PRO-FAM® NSI value on texture profile analysis and moisture/solids ratio values of dry, rehydrated and retorted textured half-products is shown in Table 4. Hardness and total compression shear (CS) force values are presented for the three physical states--dry, rehydrated, and retorted--textured soy isolates with different NSI values. Hardness and compression-shear force values for dry textured half-products increase significantly as NSI value decreases.

TABLE 4. Influence of NSI on Texture Profile Analysis Values and Moisture/Solids Ratios of Dry, Rehydrated, and Retorted PRO-FAM® Half-Products.

Variable	PRO-FAM® S970	S640	S955
NSI →	(85)	(55)	(20)
Dry[1]			
Hardness	86.63	133.22	344.34
Total C.S. Force	1.43	2.85	1.89
Moisture/Solids[2]	5.5	4.5	2.5
Rehydrated[1]			
Hardness	53.28	600.14	605.39
Total C.S. Force	1.81	7.13	6.97
Moisture/Solids	10.1	3.4	2.1
Retorted[1]			
Hardness	3.53	359.00	217.06
Total C.S. Force	0.17	4.58	3.79
Moisture/Solids	9.7	7.1	4.9

[1] 2.54 × 2.54 cm piece size.
[2] < 12 U.S. std. mesh screen particle size.

When rehydrated, hardness shows a ten-fold increase as the NSI value decreases from 85 to 55 in the starting soy isolate flour; no further change is seen as the NSI value decreases to 20. Similarly, the compression-shear force values increase, as the NSI decreases, following the same trend as the hardness values. Of interest is the change in moisture/solids ratio; as NSI decreases, there is a dramatic decrease in this ratio moving from the highly functional PRO-FAM® S-970 to the intermediate isolate, S-640. Once retorted, the same trends in hardness and compression-shear force values are observed; the most significant change occurs between the highest and intermediate functional isolates, S-970 and S-640, with NSI values of 85 and 55, respectively. The highest hardness and compression-shear force values were recorded for the retorted S-640 half-product; the reason for this cannot be explained at this time. A noticeable change in moisture/solids ratio (when compared with the rehydrated samples) is observed in retorted materials. A progressive decrease in the moisture/solids ratio occurs as the NSI values decrease. This suggests that the moisture holding capacity of the intermediate (S-640) and low functional (S-955) isolates increase as a result of retorting.

The choice of PRO-FAM® S-955 for further development of half-products using isolate/rice flour blends was by design. PRO-FAM® S-955 availability and comparable performance to the intermediate functional isolate, S-640, suggested it be used in subsequent formulated blend performance evaluations. Further, preliminary studies on the individual isolates showed that during retorting, S-955 exhibited the most stable physical shape. Integrity of piece was excellent and loss of solids to the water during the retorting step was minimal. These results suggest that PRO-FAM® S-955 is the most versatile isolate for varied half-product analog development and subsequent food product uses.

TABLE 5. Influence of PRO-FAM® Level on Textured Profile Analysis Values and Moisture/Solids Ratios of Dry, Rehydrated and Retorted PRO-FAM®/Rice Flour Textured Half-Products.

	PRO-FAM® S955/Rice Flour Blends (% Isolate)			
Variables	100	59	16	0
Dry[1]				
Hardness	134.20	599.01	N.A.[2]	559.78
Total C.S. Force	0.64	3.56	N.A.[2]	3.01
Moisture/Solids[3]	6.0	4.0	2.5	1.0
Rehydrated[1]				
Hardness	463.63	2.98	6.91	1.56
Total C.S. Force	5.15	0.14	0.33	0.12
Moisture/Solids	3.5	6.9	5.4	4.9
Retorted[1]				
Hardness	179.29	5.62	7.71	1.03
Total C.S. Force	3.20	0.49	0.20	0.06
Moisture/Solids	6.5	3.7	3.6	3.2

[1] 2.54 × 2.54 cm piece size.
[2] Not available—data off scale.
[3] $<$12 U.S. std. mesh screen particle size.

Opportunities exist to vary final textured half-product functionality and performance by selection and inclusion of carbohydrate materials. The data in Table 5 show the influence of PRO-FAM®, rice flour, and PRO-FAM® rice/flour blends on texture profile analysis and moisture/solids ratios of dry, rehydrated and retorted half-products. The data show, moving from left to right, a decrease in the percent of soy protein isolates, from 100% to 0%, in the blends. Resulting hardness and compression-shear force values for dry half-products increase as the level of soy isolate decreases. Similarly, the water-holding capacity (moisture/solids ratio) decreases as the soy isolate level decreases in the blend. These data suggest that the inclusion of rice flour increases the hardness and reduces the moisture holding capacity of a resulting dry, textured half-product blend. The resulting hardness and compression-shear force values (16% isolate level) which could not be recorded were not expected. This phenomena cannot be explained at this time. Once rehydrated, hardness and compression-shear force values for the 2.54 x 2.54 cm. piece decrease dramatically as the level of soy isolate in the blend is decreased from 100 to 59, and on to 0%. The moisture/solids ratio stabilizes at values between 3.5 and 5.4 units among all compositions. Similar trends are observed as individual components and blends of S-955 and rice flour are retorted. Hardness and compression-shear force values follow significant downward trends as the soy isolate level decreases. Similarly, the moisture/solids ratio exhibits a downward trend as the isolate level decreases in the retorted half-product.

TABLE 6. Typical Texture Profile Analysis for Some Prepared Foods.

Variable	RTE Cereal	Canned Tuna	Roast Beef
Hardness	98.18	314.75	189.15
Total C.S. Force	1.14	2.58	5.32

It should be noted that hardness and compression-shear force values for 100% PRO-FAM® S-955 reported in Tables 4 and 5 are not in agreement; the data reported in Table 5 were obtained using different HTST twin-screw parameters. This was necessitated by the different HTST extruder performance characteristics (not reported) of the formulated blends of PRO-FAM® and rice flour.

Having established the textural characteristics of HTST extruder-processed soy isolates or soy isolate/rice flour blends, how can we relate them to a finished food product? The data in Table 6 report different texture profile analysis values for some prepared foods. The texture profile analysis values for half-product analogs reported in Table 5 are not close to those for real products in Table 6; however, the opportunity exists for fabricating half-product analogs with not only texture profile, but also performance characteristics which match those of real products. Formulation and HTST twin-screw extrusion are tandem tools required to make matching or compatible half-product analogs a reality. Typical values for hardness and total compression-shear force for a ready-to-eat (RTE) cereal, canned tuna, and roast beef slices are presented. Control of HTST extrusion processing parameters can be used to modify a half-product analog to similate that of a finished food product.

In summary, we have presented a study on the use of soy protein isolates and high temperature, short time (HTST), twin-screw extrusion to prepare high protein, half-product analogs. An attempt was made to identify formulation and process parameters which contribute to significant half-product textural and performance characteristics. The possibility of matching half-product analog characteristics to those of real products is suggested.

V. ACKNOWLEDGMENT

The authors offer their gratitude to Ms. Melanie Kenyon, Food Scientist, Grain Processing Corporation, for her conscientious generation of the Instron texture profile analysis data reported in this paper.

RECENT DEVELOPMENTS IN THE APPLICATION OF XANTHAN GUM IN FOOD SYSTEMS

Gail D. Teague
Ross C. Clark
Daniel R. Burgum

Kelco
Division of Merck & Co., Inc.
San Diego, California

I. INTRODUCTION

Four recent major developments in the application of xanthan gum in food systems are:

A. Cakes - both dry mix and regular

B. Candy - starch jelly products

C. Retort pouch products and technology

D. Batters and Coatings - of the type used for seafood or poultry.

In this paper, the greatest emphasis is on cakes since considerable data are now available on the use of xanthan gum in this particular application. Before describing any of these specific applications in detail, a brief review of the basic properties of xanthan gum is necessary.

Xanthan gum, the extracellular polysaccharide from *Xanthomonas campestris*, is a food hydrocolloid having unique functional properties (1-5). These functional properties, currently utilized in a wide array of food products such as salad dressings, pie fillings, icings, toppings, sauces, and beverages, are as follows:

ISBN 0-12-169080-6

1) High at-rest viscosity, even at very low gum concentrations.

2) High yield value, and therefore good suspension.

3) High pseudoplasticity, i.e., solutions are highly shear thinning.

4) Stable viscosity in the presence of salt.

5) Retention of viscosity at elevated temperature, in the presence of low levels of salt.

6) Stable viscosity over a wide pH range.

7) Thermoreversible gelation upon heating and cooling with locust bean gum.

8) Synergistic viscosity increase with guar gum.

In many food applications the functional properties of xanthan gum are readily understood from an awareness of the aforementioned solution properties (6). However, in some complex systems, such as farinaceous foods, the utility of xanthan gum is not so readily apparent. A particularly good example is cakes, where not only are xanthan gum's solution properties important but also the interactions which occur between the xanthan gum and other ingredients, notably starch or protein.

II. XANTHAN GUM IN CAKES

In recent research studies both dry mix and regular bakery-type cake formulations were tested and the influence of xanthan gum on moisture retention, volume, ingredient and processing tolerance, batter stabilization, and overall quality of the finished product was examined.

From these laboratory studies and an awareness of existing applications of xanthan gum in the baking industry, the following benefits can be realized by using xanthan gum in cakes:

1. Increased moisture binding and retention during mixing and baking, and in finished cakes in the distribution cycle.

2. Improved volume and structure through the interaction of xanthan gum with starch and protein, resulting in less fragility in the finished product.

3. Rheology control in batters during mixing, pumping, and baking.

4. Foam or air cell stabilization during mixing, for batter uniformity and improved cell structure in the finished cake.

5. Suspension of particles such as fruit or nuts during processing and the critical early stages of baking.

6. Tolerance to variations in production conditions and ingredient quality.

7. Maintenance of good eating quality in the finished product, initially and after storage.

These seven benefits will now be discussed in more detail.

A. The Effect of Xanthan Gum on Moisture Binding and Retention

Xanthan gum, when blended with the other dry ingredients used for cake mix manufacture, hydrates rapidly and evenly to aid in preventing lumping during the critical initial mixing stage. This even hydration aids in the uniform distribution of moisture in the batters, which in turn helps stabilize the fine air cells formed during the mixing process. The stabilization of the air cells improves volume and symmetry in the finished cake.

Laboratory studies comparing cake batters prepared with and without xanthan gum showed that the batters formulated with xanthan gum were smoother and less lumpy. This difference was especially noticeable during the preparation of high moisture cakes, which are prepared by adding the water in one step.

Several cake formulations prepared in the laboratory were tested to determine the effect of xanthan gum on moisture control during the baking stage. Results showed (Table I) that the addition of only 0.05% xanthan gum (total batter weight) to the mix with no other formulation changes helped to retain moisture during baking.

TABLE I. Finished Cakes: Initial Moisture Analyses

(Vacuum drying method)	With Xanthan gum	Without Xanthan gum
Yellow cake	26.8%	25.2%
Devil's Food cake	28.5%	27.2%
Angel Food cake	35.3%	34.3%

These laboratory findings have been supported by reports from bakers and mix manufacturers who claim that cakes containing xanthan gum appear and taste more moist not only when fresh but particularly after storage for two days.

Overall quality of finished cakes is superior as a result of the moisture control provided by xanthan gum. A finer cell structure, improved volume, symmetry, and uniform texture and grain can result, as in the high moisture chocolate layer cake shown in Figure 1.

Because of the moisture control obtained with xanthan gum, it is now being used by many dry mix manufacturers to produce single stage mix products, especially retail pudding cake mixes and wholesale layer cake mixes.

B. The Interaction of Xanthan Gum with Starch and Protein to Improve Volume and Structure

Structure is usually imparted to a cake by the addition of eggs, starch and non-fat dry milk. For cost reasons, many current cake formulations contain reduced levels of these relatively expensive ingredients and, consequently, other ingredients must be added to achieve an acceptable structure. Emulsifiers and starch significantly improve the structure of these cost-reduced, high moisture cakes but, if used at too high a level, they produce undesirable finished cake qualities such as dryness, pastiness or fragility.

Xanthan gum can also be used to improve the structure of finished cakes. These improvements may be due to interactions of xanthan gum with protein and starch (7-10). The interaction with starch has been demonstrated by amylograph studies which show that the viscosity/temperature profile of a starch during cooking can be modified by the inclusion of low levels

FIGURE 1. High moisture chocolate layer cake with and without xanthan gum.

of xanthan gum. Specifically, the addition of xanthan gum lowers the temperature at which the initial viscosity increase occurs and increases peak viscosity. Such interactions are believed to occur during mixing, baking, and in the finished cake, especially after storage.

It is also known that xanthan gum can interact with gelatin, egg albumen, milk protein and wheat gluten. Interactions between xanthan gum and the proteins in the cake mix are believed to be responsible for improvements such as the increased resilience in cake crumb structure.

To test for possible interactions between xanthan gum and cake ingredients, amylographs were obtained using a bleached cake flour slurry and a diluted cake batter with and without xanthan gum. The results, shown in Figures 2 and 3, clearly indicate that xanthan gum influences the changes in viscosity which occur when these materials are tested. The effects are particularly pronounced with the cake batter.

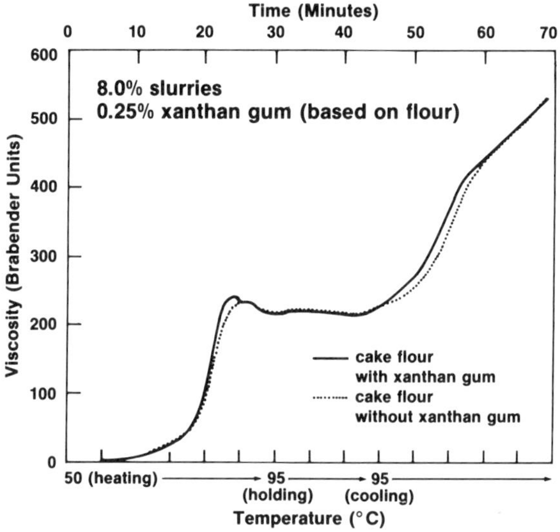

FIGURE 2. Amylographs of 8% cake batter slurries of cake flour with and without xanthan gum.

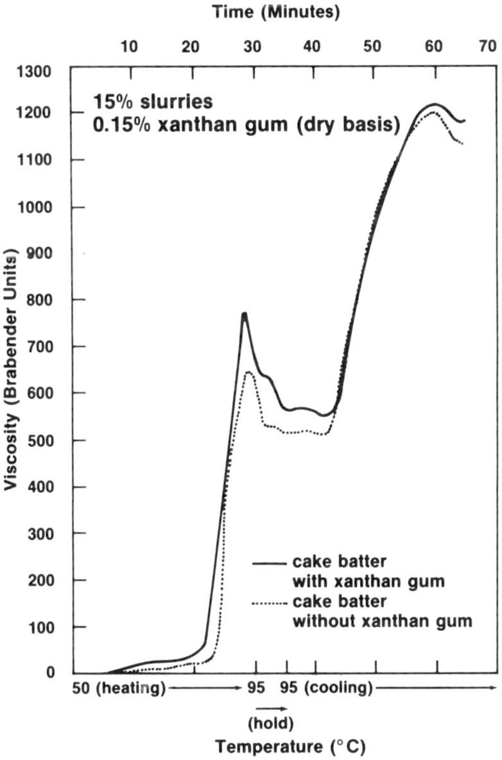

FIGURE 3. Amylographs of 15% cake batter slurries with and without xanthan gum.

A lean yellow cake formula (Table II) was used to examine the structural differences in cakes prepared with and without xanthan gum. This particular formula was selected because of the difficulties in making quality lean cakes with low egg, shortening, and nonfat dry milk levels.

TABLE II. Lean Yellow Cake Formula

Ingredients[a]	With Xanthan Gum		
	grams	%	Baker's %
Cake flour	500	25.80	100.0
Granulated sugar	600	30.96	122.0
Dried whole eggs	30	1.55	6.0
Nonfat milk solids	30	1.55	6.0
Baking powder	30	1.55	6.0
Salt	15	.77	3.0
Xanthan gum	3	.16	0.6
Shortening	100	5.16	20.0
Emulsifier	10	.52	2.0
Water	300	15.48	60.0
Water	150	7.73	30.0
Water	170	8.77	34.0
	1938.0	100.00	

[a]Yellow color added as desired

Ingredients[a]	Control		
	grams	%	Baker's %
Cake flour	500	25.70	100.0
Granulated sugar	610	31.40	122.0
Dried whole eggs	30	1.54	6.0
Nonfat milk solids	30	1.54	6.0
Baking powder	30	1.54	6.0
Salt	15	.77	3.0
Xanthan gum	--	--	--
Shortening	100	5.10	20.0
Emulsifier	10	.51	2.0
Water	300	15.40	60.0
Water	150	7.70	30.0
Water	170	8.80	34.0
	1945.0	100.00	

[a]Yellow color added as desired

The structural quality of the cakes was measured objectively using an Instron Model 1122 Universal Tester. The force required to compress a fixed volume of cake (1 cubic inch) to half of its original height was used as a measure of cake firmness. The cake sample was subsequently held in the compressed state and the rate of decay of the applied force measured. For elastic structures, this rate of decay is slow; for more fluid, less elastic structures the rate of decay is more rapid. The rate of decay is thus a good indicator of a cake's consistency. Changes in applied force with time were continuously monitored as a trace on the instrument chart recorder.

Results obtained with the lean cake with and without xanthan gum are shown in Figure 4.

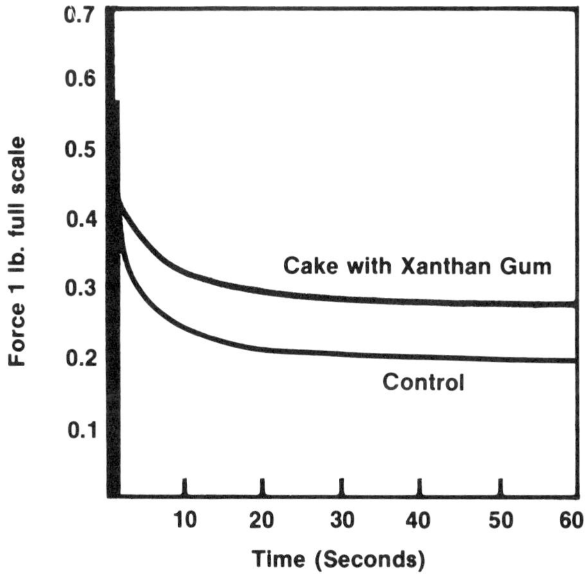

FIGURE 4. Instron texture measurements of lean yellow cake with and without xanthan gum.

The xanthan gum curve is shifted upward but retains the same shape as the control curve. This indicates that xanthan gum has increased the firmness without changing the other properties of the cake. In other words, although firmness is increased, other eating qualities remain the same. Because of the added firmness, the handling qualities are improved. This is especially important in high moisture systems, which are noted for being fragile. When xanthan gum is added to these systems, they are strengthened without becoming undesirably tough. In this way, xanthan gum can reduce the degree of cracking or crushing during distribution.

The texture and grain of the crumb of the lean yellow cake were photographed to illustrate the positive effect obtained with the xanthan gum, which is readily apparent in Figures 5 and 6. The cake with xanthan gum was found to cut more cleanly, which is an indication of improved crumb strength.

FIGURE 5. Lean yellow cake with and without xanthan gum

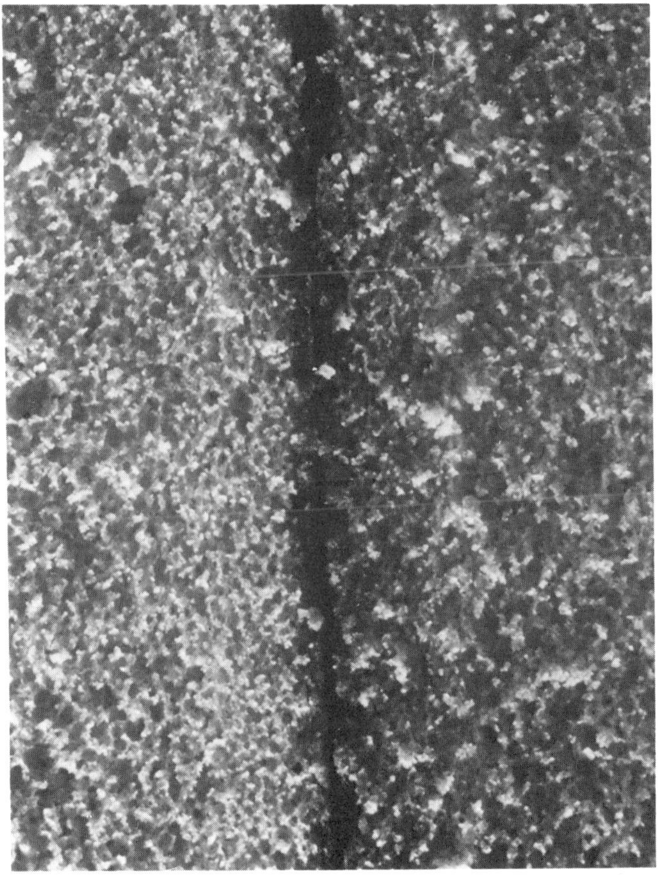

FIGURE 6. Lean yellow cake cell structure with (left) and without (right) xanthan gum.

The structural benefits obtained with xanthan gum are particularly useful for high moisture cakes formulated with corn or tapioca starch. Strength is added to these cakes without any toughening of the crumb. The cakes also appear to be fresher as a result of modification of the starch retrogradation process by xanthan gum.

Although the ability of xanthan gum to improve volume in high moisture cakes is related to control of moisture in the system, it is also related to the structural improvement attained by the protein and starch interactions. Laboratory studies have shown that cakes can increase 10% in volume because of the moisture control and ingredient interaction provided by xanthan gum. Field studies have confirmed these improvements with xanthan gum, and reports indicate 10-15% volume increases in cakes prepared from dry mixes and in finished baked cakes. Figure 7 shows the 10% volume increase attained with xanthan gum added at 0.15% (flour basis) to a chocolate pudding cake prepared in the laboratory.

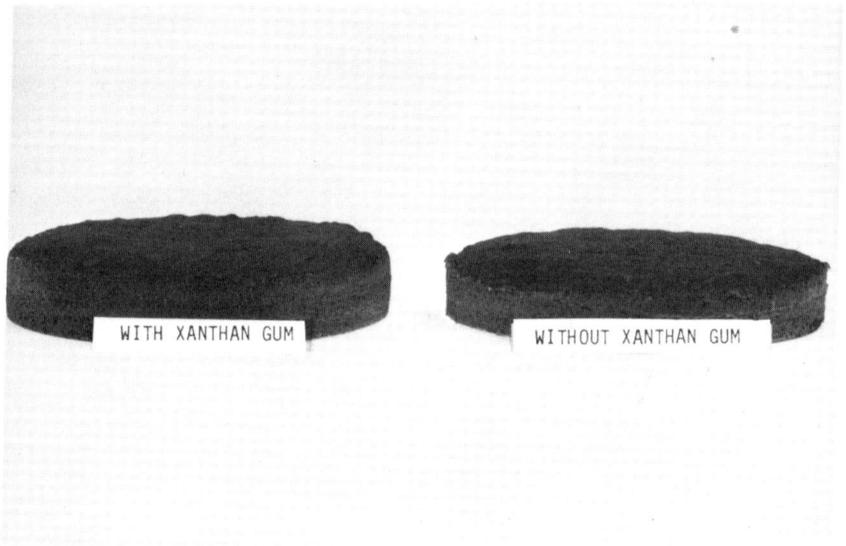

FIGURE 7. Chocolate pudding cake with and without xanthan gum.

C. The Effect of Xanthan Gum on Batter Rheology During Mixing, Pumping and Baking

When xanthan gum is added to aqueous systems a highly significant change in rheology occurs. The systems become more viscous and also more pseudoplastic. Pseudoplastic systems have very high viscosities at low shear rates but as the shear rate increases (during processing, for example) the viscosity diminishes as shown in Figure 8. As soon as the shear is removed, however, viscosity is recovered.

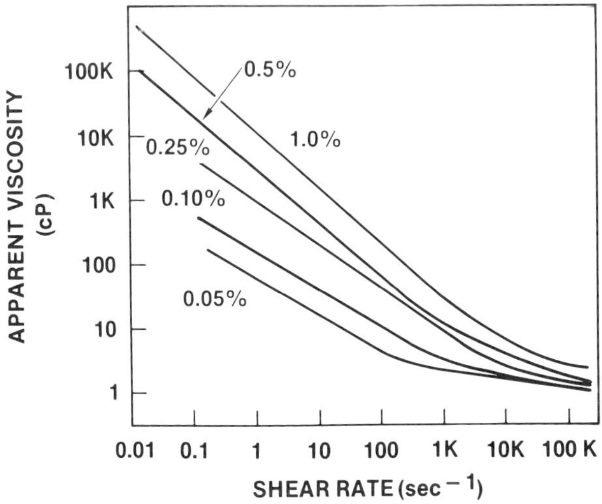

FIGURE 8. Xanthan gum: Viscosity versus shear rate.

Cake batters benefit from this pseudoplasticity in several ways. Mixing, pumping and filling are facilitated because at the high shear rates encountered in these operations the batters containing xanthan gum have a low viscosity. When the batter from the filling head comes to rest in the pan the original high viscosity returns and splashing is minimized. Batters which have a low viscosity tend to splatter on contacting the hard surface of the pan, causing problems during filling. The high low-shear-rate viscosity which xanthan gum imparts to batters can also increase air retention resulting in batters with low specific gravities which produce lighter cakes. The influence of xanthan gum on air incorporation is further discussed in the following section.

D. The Effect of Xanthan Gum on Foam or Air Cell Stabilization During Mixing and Baking

Xanthan gum has been found to be very effective in stabilizing air cells in cakes. Although it can be used for this purpose in any cake, it is especially beneficial in highly aerated systems. Laboratory experiments were conducted to quantitatively measure the foam stabilizing effect of xanthan gum in a chocolate layer cake, a medium rich angel food cake (prepared with frozen egg white), and a snack type yellow sponge cake. Quantitative measurements on the cell structure were made on the finished cake by photographing a freshly cut surface covered with a transparent 5 mm grid using a 35 mm camera and counting the intact air cells visible on a projected photograph.

Figures 9 and 10 illustrate the type of air cell distribution obtained in a dark chocolate cake. The number of air cells increased approximately 10% or 1 cell per $5(mm)^2$ with the addition of 0.05% xanthan gum (total batter weight).

A second series of experiments evaluating air cell stabilization were conducted to assess the volume improvements obtained by the addition of xanthan gum to a cake mix. Yellow sponge and angel food cakes were examined. The yellow sponge cake with xanthan gum is compared to the control with no xanthan gum in Figure 11. Part of the 15% volume improvement is believed to be due to the better emulsion stabilization provided by xanthan gum.

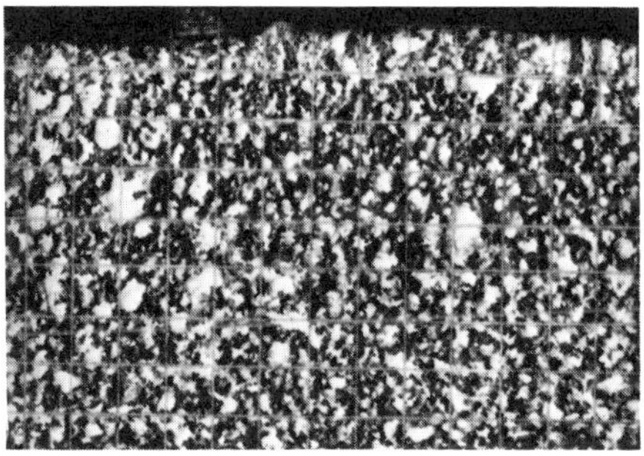

FIGURE 9. Objective measurement of cake cell structure - dark chocolate cake without xanthan gum.

FIGURE 10. Objective measurement of cake cell structure - dark chocolate cake with 0.05 percent xanthan gum (total batter weight).

FIGURE 11. Snack type sponge cake shows improved overall volume and symmetry with addition of xanthan gum (no differrence in added water).

Angel food cakes require egg albumen to achieve highly foamed batters, which result in a delicate, aerated, fine-cell-structured finished cake of high volume. Because of the ability of xanthan gum to stabilize foam or air cells, experiments were conducted to determine the possibility of reducing the amount of frozen egg white required. It was found that up to 20% of the egg white in a medium rich angel food cake could be replaced by the addition of approximately 0.05% xanthan gum without adversely affecting the finished product quality. Adjustments in the level of added water are necessary to compensate for the lowered frozen egg white level, since frozen egg white has a high water content.

E. The Effect of Xanthan Gum on Suspension of Particles Such As Fruits or Nuts

Suspension of nuts and/or fruit particles in batters during the production of cakes and during dry cake mix preparation in the home is an important factor in attaining a uniform appearance. Suspension of these ingredients is particularly difficult with cake mixes in which all the liquids are added in the cake pan (e.g. one stage snack-type cakes) with a minimum amount of mixing, as with a fork.

The ability of a fluid to suspend particles depends, among other factors, on the viscosity of the fluid. The higher the viscosity the less is the tendency for the suspended particles to settle. Clearly, additives such as xanthan gum which impart high at rest viscosity greatly improve suspension, and xanthan gum is frequently used in batters to provide improved suspension of fruit or nut pieces, resulting in a more appealing cake. As indicated in the introduction, the viscosity of xanthan gum is stable over a wide temperature range. Xanthan gum thus helps to maintain good suspension stability while the cake batter is being heated during the initial stages of the baking process. In the reconstitution of dry cake mixes, the ability of xanthan gum to hydrate rapidly greatly improves the uniform suspension of particulates. The high viscosity under low shear conditions and the high yield value obtained with xanthan gum (See Figure 8) reflect its excellent suspending ability.

Xanthan gum is currently being used in a number of dry mix cakes for its excellent suspending properties, even when added at low concentrations, e.g. 0.05 - 0.20 percent (based on total dry weight).

F. The Effect of Xanthan Gum Upon Operating and Recipe Tolerance

The inaccurate measurement by the consumer of the amount of water added to a retail cake mix is a common occurence. Similar errors can occur when measuring water or other ingredients in the preparation of institutional cakes or during the large scale production of finished cakes. Cake formulations which have a built-in tolerance to such errors and which accommodate unavoidable variations in production conditions and in the functionalities of the basic ingredients such as flour and soy protein are highly desirable. Recipe tolerance is particularly difficult to achieve with "lean" cakes which contain relatively low levels of egg, non-fat dry milk, and shortening. They are sensitive to even minor changes in ingredient quality or make-up procedure. Cake formulations of this type are often said to be "stressed". Xanthan gum has been found to provide excellent recipe tolerance in these stressed cake formulations.

Laboratory tests were conducted to demonstrate the degree of recipe tolerance (or operating tolerance) possible with xanthan gum in the lean cake formulation shown in Table II. Figure 12 compares the finished cake containing 0.14% xanthan gum (based on total batter weight) prepared with 20% additional water to the control cake with 20% additional water and no xanthan gum. The rings in the cake without xanthan gum indicate that the flow of the batter was not optimal with the high water level. The recipe tolerance which xanthan gum imparted to the cake with 20% additional water almost completely eliminated the ringing and control of the flow properties of the batter was achieved. Also, the control cake containing 20% more water had a pasty mouthfeel while the same cake with xanthan gum had a significantly improved eating quality.

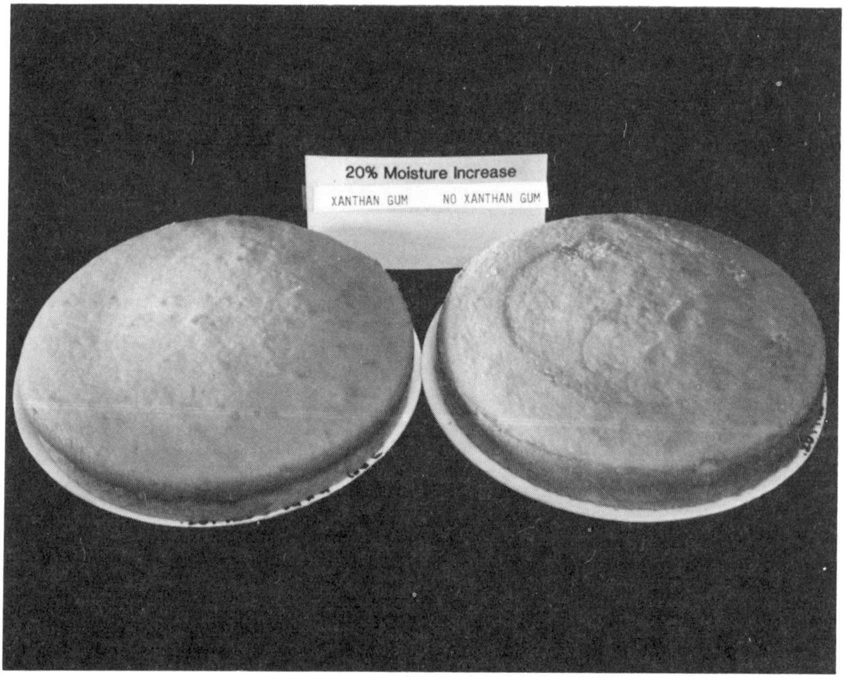

FIGURE 12. Lean yellow cake with 20% additional water with and without xanthan gum.

The recipe tolerance with added xanthan gum was also tested in the lean cake formula prepared with a 25% reduction in the amount of added water. The dramatic difference in volume, structure, and surface appearance of the cake with 0.10% (total batter weight basis) xanthan gum compared to the control is shown in Figures 13 and 14.

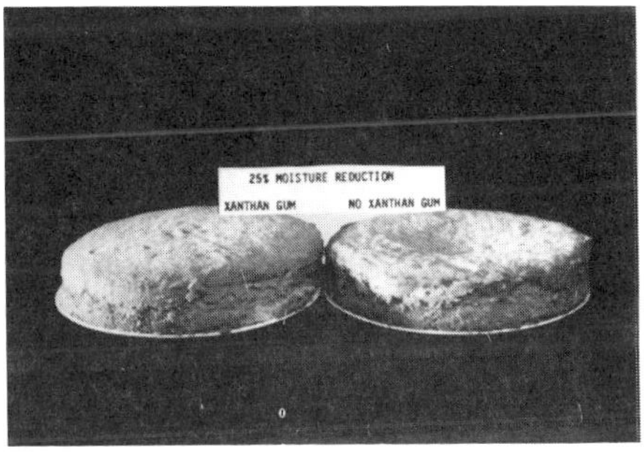

FIGURE 13. Lean yellow cake with 25% reduction in water with and without xanthan gum.

FIGURE 14. Sliced lean yellow cake with 25% reduction in water with and without xanthan gum.

Other researchers have found that specialty products such as reduced calorie, dietary, and low cholesterol cakes, which fall into the "stressed ingredient" category and require a high degree of recipe tolerance, have this character imparted to them by the inclusion of xanthan gum (11). This suggests that xanthan gum is effective not only in minimizing variations in the amounts of added water but also in masking differences in the functionality of the other ingredients.

G. The Effect of Xanthan Gum upon the Eating Quality of Finished Cakes

Cakes with improved eating quality are the net result of attaining the benefits described. Sensory evaluations conducted in the laboratory correlate with those reported in the field which indicate that cakes containing xanthan gum are more moist initially and especially after being stored for two days. The texture of the cakes both in terms of appearance and organoleptic evaluation is more pleasing, i.e., less cracking is evident, the cake is more symmetrical, the texture is uniform and the mouthfeel less pasty with a silkier crumb.

It should be pointed out that all of the benefits described are not apparent in every cake application, especially if the usage level is not optimized. To obtain the best results with xanthan gum, the usage level should be in the range of 2-3 ounces per 100 pounds of flour or 0.03-0.20% (based on total batter weight) depending upon the levels of other ingredients.

III. XANTHAN GUM IN CANDY

Recent developments in the application of xanthan gum in starch jelly candy utilize the reaction between xanthan and locust bean gum (12). When xanthan gum is combined with locust bean gum, a synergistic viscosity increase occurs (Figure 15). At a sufficiently high total gum concentration, a gel can be formed as a result of this synergism and this gel can be used to an advantage in the manufacture of starch jelly candy (13). A blend of xanthan and locust bean gum can be used in these candies to (1) accelerate gelling to reduce set time, (2) extend shelf life, (3) modify texture of the candy.

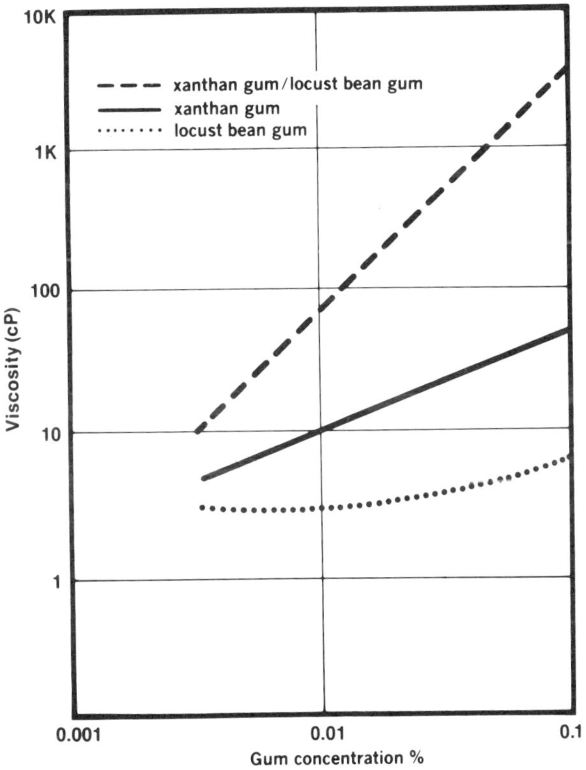

FIGURE 15. Synergistic viscosity increase with xanthan/ locust bean gum blend.

Figure 16 demonstrates the reduction of set time possible by the addition of a blend of xanthan gum/locust bean gum at approximately 0.5 percent (finished candy basis) (10). Three starch jelly systems were studied. These were:

1) A thin-boiling starch system

2) A high-amylose starch system

3) A system using a blend of thin-boiling and high amylose starches.

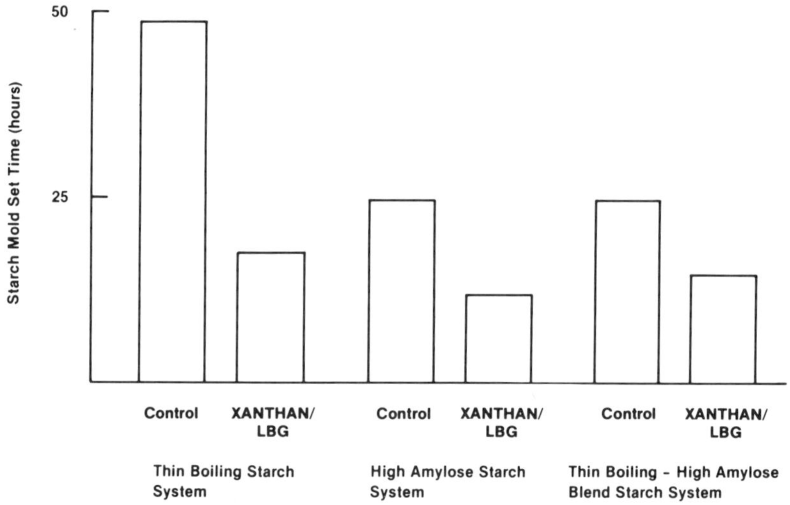

FIGURE 16. Reduction of set time with xanthan gum and fine mesh locust bean gum (Approximately 0.5%).

All three systems showed a vast reduction of set time by the addition of this gelling blend to the formula. An example of a thin boiling starch system which can be used is shown in Table III.

TABLE III. Starch Jelly Candy Prepared with Xanthan Gum/ Locust Bean Gum Blend

Ingredients	Quantity
Sugar, refined sucrose	31.90
Corn syrup, 65 DE	26.26
Corn syrup, 42 DE	18.76
Water	12.20
Thin boiling corn starch	10.32
Xanthan gum/locust bean gum	0.56
Flavor and color (to suit)	
	100.00%

As mentioned, this blend is also effective in prolonging the shelf life of the candies, by inhibiting starch retrogradation and moisture loss with time. The inclusion of the blend modifies the texture of the candies and jelly candies produced with the xanthan/locust bean gum blend are usually more tender, more moist, and less sticky compared to regular starch jellies. For effective use of the blend, proper hydration and dispersion are essential.

IV. XANTHAN GUM IN FOODS PROCESSED IN RETORT POUCHES

Another new application of xanthan gum is in foods packaged in retort pouches. The retort pouch is a relatively recent packaging form that offers unique advantages to the manufacturer and to the consumer. For example, improved food quality results from the reduced processing time required for sterilization of retort pouch foods. Also retort pouches are convenient to handle, require less reheating time, and less space for storage or disposal. The retort pouch offers challenges as well as advantages to the food manufacturer.

1) Often, use of highspeed equipment to fill the pouches compounds the problem of properly filling the retort pouch without splattering or dripping.

2) Also, as with canned products, the high temperatures required for heat processing can result in impaired product quality.

The use of xanthan gum in retort pouches helps alleviate these problems. Essentially, retort pouch foods containing xanthan gum have two benefits:

1) Pseudoplastic flow - which helps facilitate mixing and filling

2) Excellent heat penetration properties and heat stability.

As previously shown in Figure 8, at the high shear rates encountered during mixing and filling the viscosity of a food system containing xanthan gum is low and consequently less energy is required for these operations. When shear rates are reduced after filling, the viscosity increases, minimizing splashing. Control of splashing is essential to produce a properly sealed pouch (15).

As shown in Figure 17, the viscosity of solutions containing xanthan gum is quite constant as the temperature is increased. At retort temperatures (116-121°C, 240-250°F), however, the viscosity is markedly reduced. The resulting thinner solution is more conducive to convection heating, which is efficient and promotes good heat penetration. Upon cooling, the high viscosity of the solution returns to give the desired consistency.

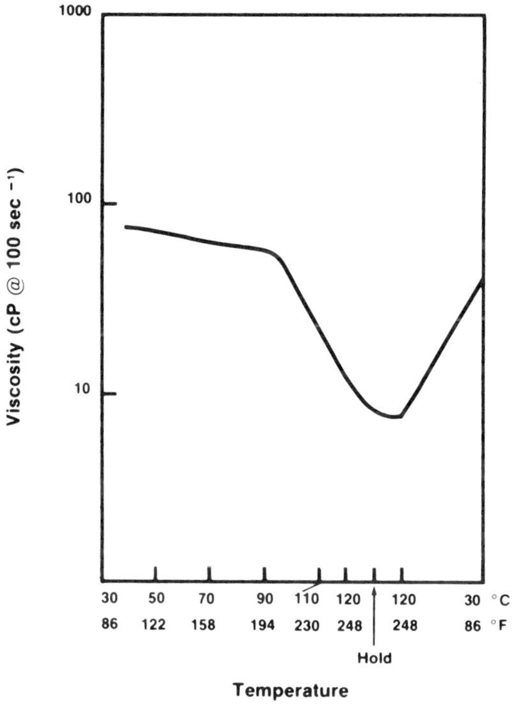

FIGURE 17. Viscosity versus Temperature - 0.5% xanthan gum in 0.5% NaCl.

V. XANTHAN GUM IN BATTER COATINGS

The final area of applications research to be described is the functionality of xanthan gum in seafood and/or poultry batters. These batters are frequently used for the large scale production of batter coated frozen fish, shrimp, or chicken. Batter viscosity loss over time is a common problem associated with the manufacture of such products. The viscosity loss causes adhesion problems and difficulties in attaining the proper texture in the final product.

Studies were carried out in the laboratory to compare the viscosity stability of batters prepared with xanthan gum to that of batters prepared with guar gum, carrageenan or locust bean gum.

The test formulation consisted of 100 parts flour, and 350 parts water with varying levels of the gums mentioned. The temperature of the batters was maintained at approximately 20°C for 3 hours under constant agitation to simulate mixing and pumping during production. Figure 18 illustrates the dramatic differences in viscosity stability of batters prepared with guar gum, locust bean gum, carrageenan, and xanthan gum.

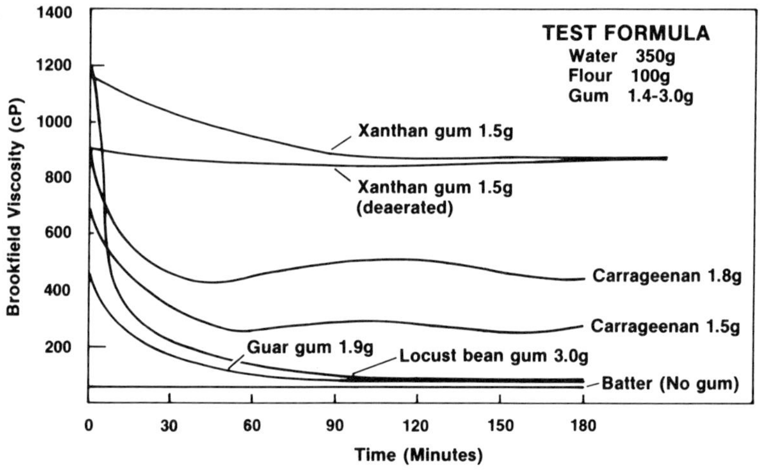

FIGURE 18. Batter coatings containing various gums.

As can be seen, xanthan gum displayed superior functionality. Xanthan gum is enzyme resistant and is not degraded by enzymes present in flour. It is also shear stable and would not be expected to lose viscosity from mixing and pumping in a commercial operation. In addition, it can be used at a lower concentration than any of the other gums tested. Xanthan gum's pseudoplastic nature allows for a more even coating and less dripping. This was evident when several pieces of parboiled chicken were dipped in the batter. The batter containing xanthan gum provided a good even coating and very little run-off was observed. The batters containing guar gum and locust bean gum left a very thin uneven coating and excessive run-off resulted. The batter containing carrageenan left an even coating, but more run-off was observed when compared to the batter containing xanthan gum.

Results of these trials could influence the choice of stabilizer for any farinaceous food application in which flour, water, and gum are used, and where the product is held for a period of time before cooking.

Acknowledgments

The authors acknowledge the technical and editorial assistance provided by Dr. George R. Sanderson.

References

1. Pettitt, D. J., in "Polysaccharides in Foods", (Blanchard, J. M. V., and Mitchell, J. R., Eds.), p. 263. Butterworths, Boston, (1979).

2. Sanderson, G.R. "Interactions of Xanthan Gum in Food Systems" presented at International Conference on Gums and Stabilizers for the Food Industry, Clwyd, (1981).

3. Sanderson, G.R., Food Technology, 35, 7, p. 50 (1981).

4. Cottrell, I.W., Kang, K.S., and Kovacs, P., in "Handbook of Water-soluble Gums and Resins," (Davidson, R.L., Ed.), pp. 24-2 - 24-16, McGraw-Hill, New York, (1980).

5. Cottrell, I.W. and Baird, J., in "Kirk-Othmer: Encyclopedia of Chemical Technology", vol. 12, 3rd ed., pp. 62-64. John Wiley, New York, (1980).

6. Sanderson, G.R., British Polymer Jnl., 13, 71 (1981).

7. Christianson, D. D. and Gardner, H. W., British Patent 1,562,275 (1980).

8. Cheng, H. and Wintersdorff, P., U. S. Patent 4,298,729 (1981).

9. Shim, J. L. and McConnell, J. L., "Interaction Between Corn Starch and Xanthan Gum" (to be published).

10. Singer, N. S. and Murray, D. W., U. S. Patent 4,198,438 (1980).

11. Bakery Production and Marketing, 15, 91 (1980).

12. Kovacs, P., Food Technology, 27, 26 (1973).

13. Cheng, H., U. S. Patent 4,219,582 (1980).

14. "KELTROL® Xanthan Gum With Locust Bean Gum in Starch Jelly Candies", Technical Bulletin F-80, Kelco, Division of Merck & Co., Inc., San Diego, (1981).

15. "KELTROL® Xanthan Gum for Use in Foods Packaged in Retort Pouches", Technical Bulletin F-79, Kelco, Division of Merck & Co., Inc., San Diego, (1981).

POLYDEXTROSE - A NEW FOOD INGREDIENT: TECHNICAL ASPECTS

Robert P. Allingham

Bio-Organic Chemicals Research and Development Department
Pfizer Inc.
Groton, Connecticut

I. INTRODUCTION

Most people think of sugar in terms of its sweet taste and the pleasure obtained from eating foods containing it, such as ice cream and cake. However, sugar also provides bulk, texture, and mouthfeel without which the preparation of these foods would be difficult or impossible. Unfortunately, for diet-conscious people, the pleasure of eating sugar in foods is always accompanied by sugar's calories.

The need for a low calorie replacement for sugar arises from the fact that many people consume too many calories, and sugar is an important contributor to this caloric excess. Heavy consumption of sugar is often associated with overweight, which is one of the most common and serious nutritional problems in this country. For example, nearly 20% of American adults are overweight to a degree that may interfere in some way with health, and nearly 15 million are at serious risk to health for this reason (Nelson, 1973; Actuaries, 1959). Overweight and obesity are often cited as predisposing factors to hypertension, and to diabetes which afflicts about 5 million Americans (Mayer, 1969, 1972; Dwyer, 1972). Dental caries affect nearly the entire U.S. population, and consumption of sucrose is said to be the most important single contributing factor (Yudkin, 1972).

Clearly, reduced consumption of sugar and its calories could make a useful contribution toward preventing problems like these by helping people achieve and maintain reasonable body weight consistent with good health (Krall, 1971; MacDonald, 1971; McGandy, 1967). Unfortunately, even with these and other compelling warnings, the success rate for keeping to diets for obesity or diabetes is discouragingly low. It is not easy to break established habits of eating too much, especially of foods eaten for pleasure, such as sweets (Sassoon, 1973).

ISBN 0-12-169080-6

For these reasons we began a search for a low calorie sugar substitute which would be useful in sensibly designed programs for weight control. The objective was a low calorie product with high water solubility and little or no color or flavor, which could provide the bulk and mouthfeel of sugar without sweetness. Artificial sweeteners could be used to adjust sweetness as desired. Safety and cost were, of course, major considerations also.

II. PREPARATION AND PROPERTIES

Polydextrose, invented by Dr. Hans Rennhard (1973) in our Groton Research Laboratories, was the end result of an extensive synthesis program to meet this objective. Reactions of glucose and other simple carbohydrates with all the common food grade alditols and carboxylic acids under a wide variety of conditions and ratios were investigated. Optimum results were obtained with the material we now call polydextrose, which is a randomly-bonded condensation polymer of glucose containing minor amounts of bound sorbitol and citric acid; it is prepared by vacuum polycondensation of a molten mixture of the starting materials in a ratio of 89:10:1.

Analyses of this product by gel permeation chromatography and gas liquid chromatography show the typical composition outlined in Table I. This shows that both sorbitol and citric acid become covalently bonded to the polymer during the reaction.

Some typical properties of polydextrose are shown in Table II.

TABLE I. Polydextrose: Typical Composition

Polydextrose polymer	minimum	90%
Glucose	maximum	4%
Sorbitol	maximum	2%
Levoglucosan	maximum	4%
Water	maximum	4%
Citric acid (free)	approximately	0.1%

TABLE II. Polydextrose: Typical Properties

Reducing value (Hagedorn-Jensen)	5
Solubility in water at 25°C (%)	80
Non-dialyzable fraction (%)	71
Color (% T at 560 nm, 10% solution)	96
pH (10% solution)	3.2
Acid equivalent (mg NaOH/g)	2.4
Optical rotation	+60
Viscosity (cps, 50% solution)	35
Intrinsic viscosity [η]	0.03

Sorbitol plays an important role in the polymerization by helping to control the upper molecular weight limit and preventing the formation of water-insoluble material. The molecular weight distribution is illustrated by the gel permeation chromatograms shown in Figures I and II. Figure I is a Sephadex G-50 curve which emphasizes the higher molecular weight region. Note that there is little or no polymer with molecular weight over 15,000 or so. Distribution in the lower molecular weight region is shown by the Sephadex G-15 curve in Figure II. Note that 1,6-anhydroglucose separates from glucose and sorbitol. The molecular weight scales were calibrated with polydextrose markers made without sorbitol so that the number-average molecular weights of individual chromatogram fractions could be determined by the Somogyi copper phosphate method. From these data a percentage molecular weight distribution was derived, and is summarized in Table III.

TABLE III. Approximate Molecular Weight Distribution of Polydextrose

Molecular Weight Range	Percent
162 - 5,000	88.7
5,000 - 10,000	10.0
10,000 - 16,000	1.2
16,000 - 18,000	0.1

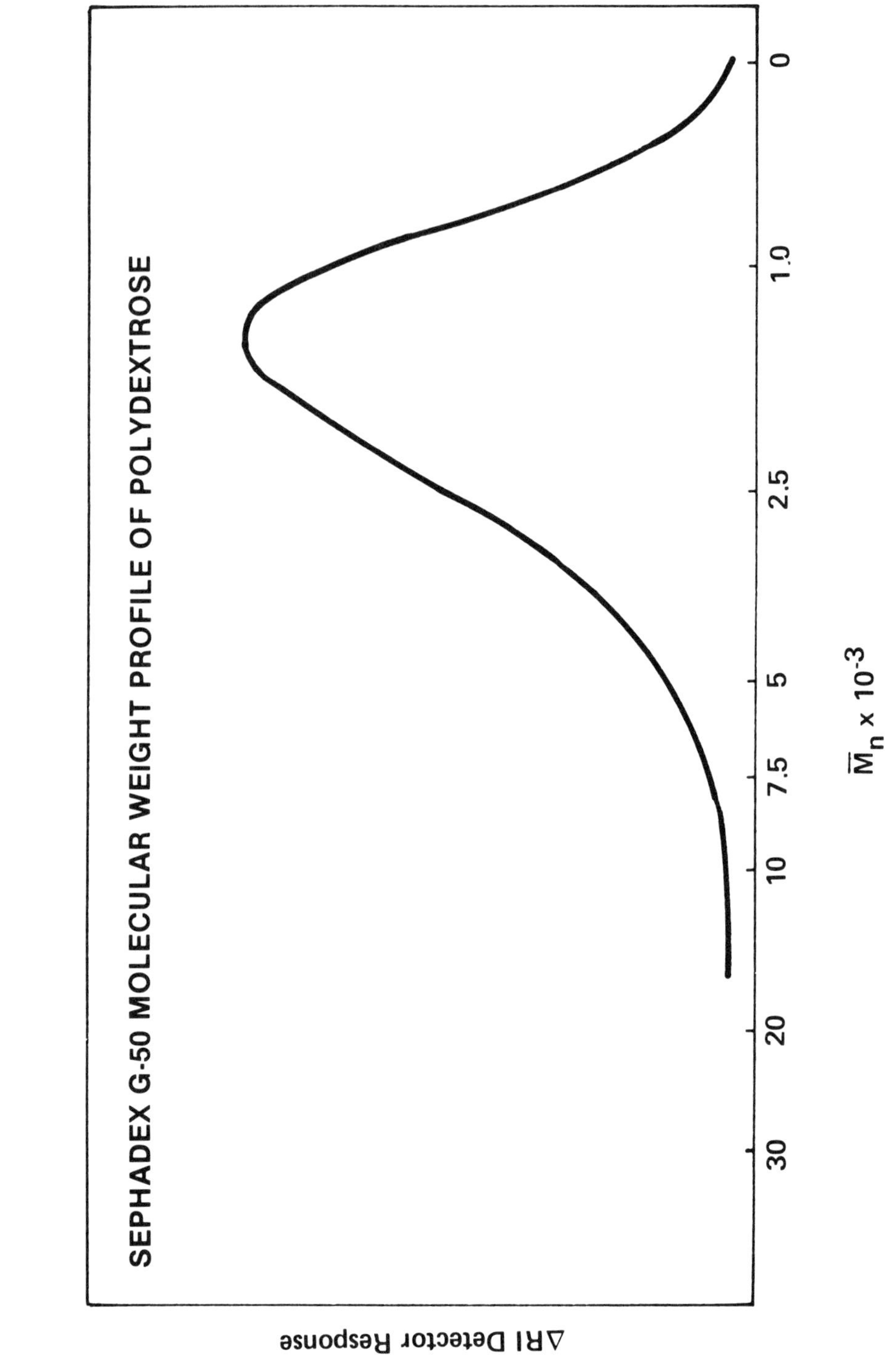
SEPHADEX G-50 MOLECULAR WEIGHT PROFILE OF POLYDEXTROSE
0
1.0
2.5
5
7.5
10
20
30
$\overline{M}_n \times 10^{-3}$
ΔRI Detector Response

Figure I

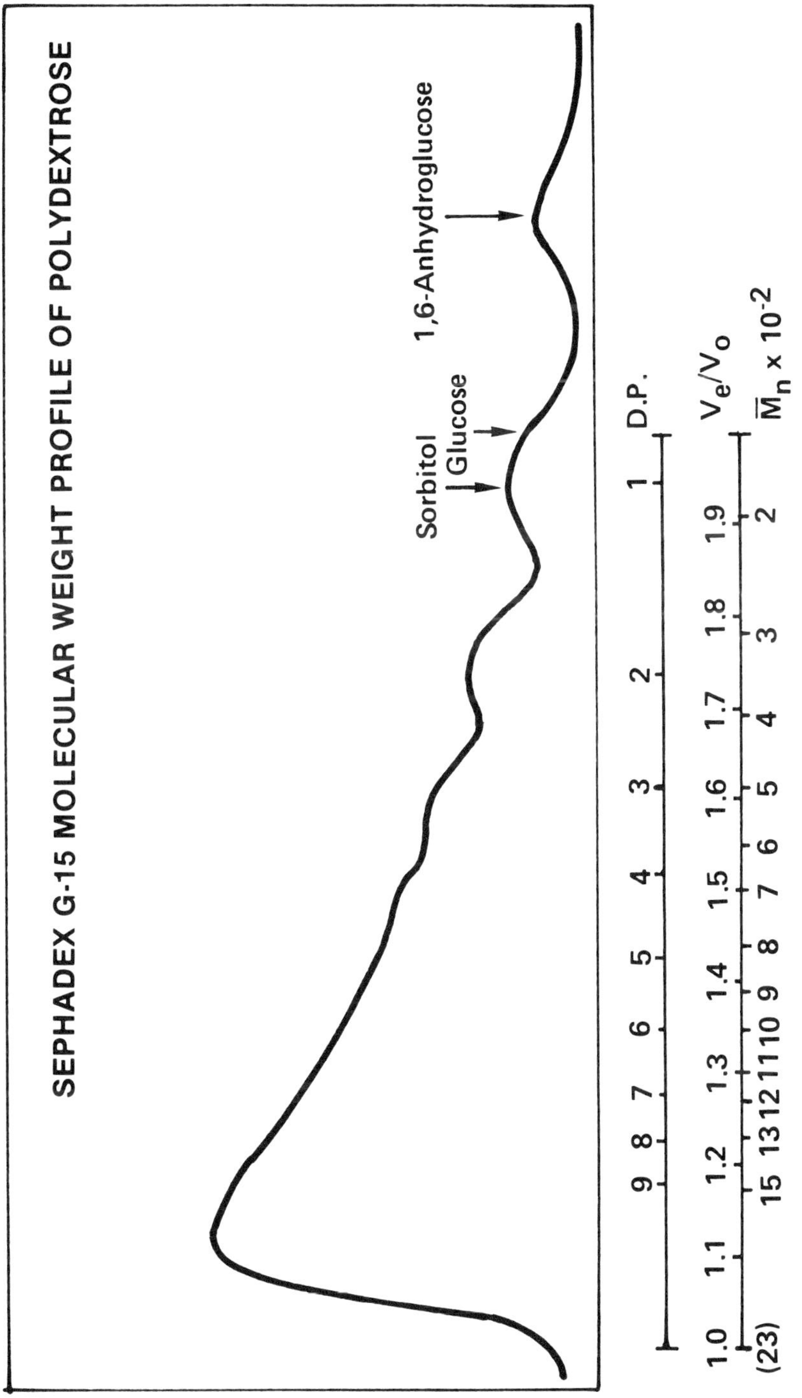
SEPHADEX G-15 MOLECULAR WEIGHT PROFILE OF POLYDEXTROSE
1,6-Anhydroglucose
Sorbitol
Glucose
D.P.
1 2 3 4 5 6 7 8 9
V_e/V_o
1.0 1.1 1.2 1.3 1.4 1.5 1.6 1.7 1.8 1.9
$\overline{M}_n \times 10^{-2}$
(23) 15 13 12 11 10 9 8 7 6 5 4 3 2

Figure II

III. STRUCTURE AND UTILIZATION

The structural features of polydextrose were determined by classical wet chemical methods, and the results are illustrated by the hypothetical part structure shown in Figure III. Major points to keep in mind are that all theoretically possible linkages are represented, including both anomeric forms and both cyclic forms of glucose, and that the molecules are highly branched. This is in direct contrast to most natural carbohydrates such as starch, which contains relatively few linkage types, much less branching and usually only one form of glucose.

The methods used in the structure work included methylation analysis and periodate oxidation to determine degree of branching; Smith periodate degradation, methylation and acetolysis to determine linkage positions, and Smith degradation and acetolysis to determine the configuration of anomeric linkages. The conclusions from this work were also confirmed by a ^{13}C-NMR study.

One might expect enzymes to have some difficulty in hydrolyzing carbohydrate molecules of such complexity, and this was found to be true in this case. The caloric utilization of polydextrose was determined by radio-tracer studies in both rats and humans to be about one-fourth of that of sucrose or about one calorie per gram (Figdor, Rennhard, 1981).

Although it is practically inert to mammalian digestive enzymes, polydextrose is partially fermented by fecal microorganisms which produce some volatile fatty acids, or VFA. The VFA are absorbed and calorically used by the host, shown by measurement of breath carbon dioxide ^{14}C, corresponding to a caloric utilization in man of no more than 27% (Pfizer, to be published). The utilization pathways are represented schematically in Figure IV.

IV. SAFETY

The safety evaluation of polydextrose included a large number of both standard and non-standard studies. This was necessary because even partial substitution for a major ingredient such as sucrose would result in ingestion of significant levels of polydextrose as opposed to normal levels of flavorants, for example. Standard toxicology studies were generally uneventful and showed that polydextrose could be included in the diet at high levels for long periods of time without harmful effects. The only untoward

HYPOTHETICAL STRUCTURE
OF POLYDEXTROSE
OR REFEATING UNIT

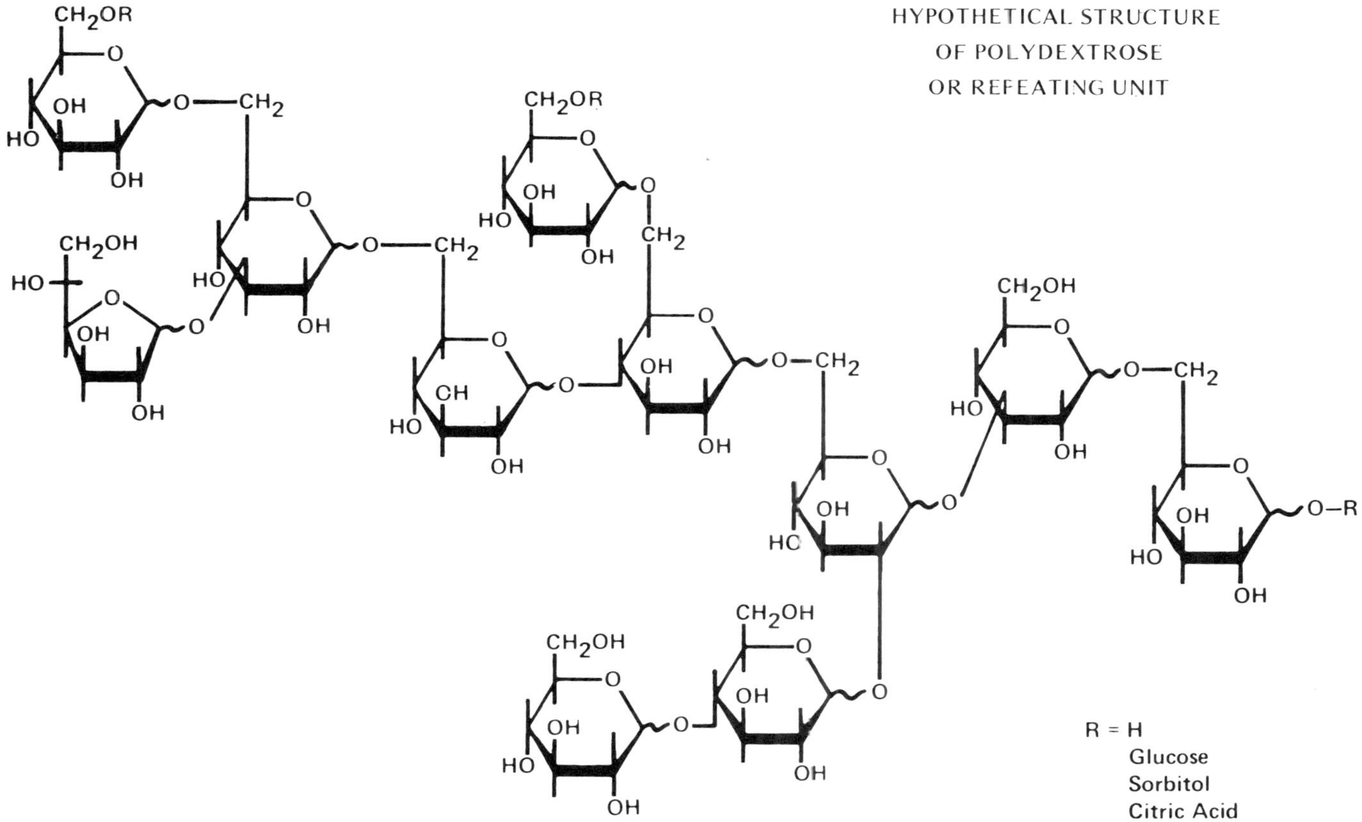

R = H
Glucose
Sorbitol
Citric Acid
Polydextrose

Figure III

POLYDEXTROSE CALORIC UTILIZATION

1 Calorie per gram

- Inert to digestive enzymes
- Large-intestine microorganisms only capable of partial conversion:

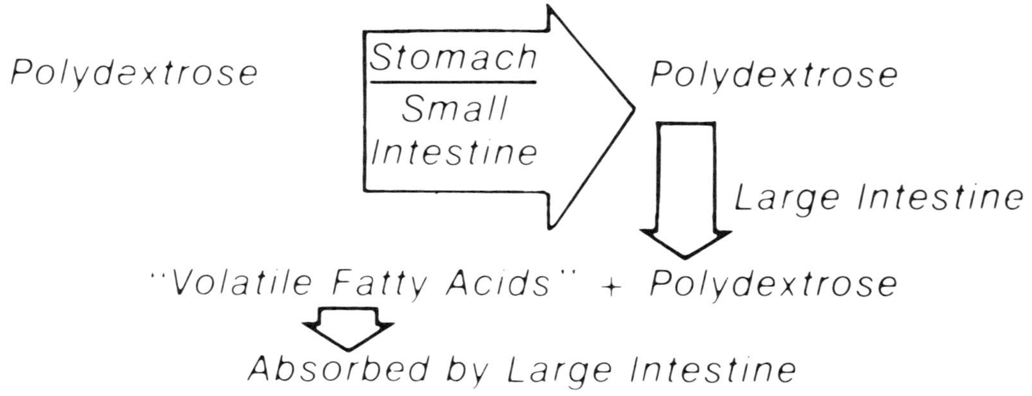

- No significant adaptation of microorganism to greater utilization

Figure IV

effect noted arose in dogs consuming excessively high doses resulting in diarrhea and its sequelae. The feeding levels used in these studies are summarized in Table IV.

The eight clinical studies were somewhat more unusual and included several hundred adults and children as subjects. These studies showed that polydextrose was well-tolerated when consumed throughout the day as an ingredient in meals or snacks. As expected, excessive consumption can lead to mild gastrointestinal effects similar to those from other poorly digested foods such as beans or sauerkraut. In a simultaneous study, polydextrose was somewhat better tolerated than sorbitol under identical conditions.

A metabolic balance study done at MIT by Professors Scrimshaw and Young showed that polydextrose does not interfere with absorption and utilization of essential vitamins, minerals, or amino acids, and a study done by Professor McMahon at Tulane University showed no significant increase in blood glucose or insulin demand for a 50-gram slug dose under glucose toleration test conditions.

Finally, studies by Professor Muhlemann at the University of Zurich showed that polydextrose has a very low cariogenicity potential.

The major conclusions from these studies are summarized in Tables V and VI.

TABLE IV. Safety Evaluation Studies: Feeding Levels

Acute toxicity (mouse, rat, dog)	>20 g/kg
Subacute toxicity (rat, monkey)	1, 2, 10 g
Reproductive studies (rat)	1, 2, 4 g/day
(rabbit)	3, 6, 12 g/day
Genetic toxicology (*in vitro* cytogenetic and mouse bone marrow chromosome studies)	
Eighteen month (mouse)	5, 10% of diet
Twenty-four month (rat) (3-generation)	5, 10% of diet
Twenty-four month (dog)	10, 20, 50% of diet

TABLE V. Clinical Studies: Conclusions

1. Well tolerated at normal ingestion levels
2. No interference with absorption and utilization of essential dietary components -- vitamins, minerals, amino acids
3. No increase in blood glucose levels in diabetics
4. Gastrointestinal effects only at high ingestion levels
5. 90 g/day average threshold for laxative effect (vs. 70 g/day for sorbitol)

TABLE VI. Cariogenicity Studies (Professor H. Muhlemann, University of Zurich)

1. Inert to *in vitro* test systems (*S. mutans*, enzymes)
2. Does not cause caries in rats under standard test conditions
3. Passes the human interproximal plaque pH test

In summary, polydextrose is one of the more complex polysaccharides and is one of the most carefully studied food ingredients ever investigated. It has a caloric utilization of one calorie per gram as shown by radio-tracer studies. The Food and Drug Administration has indicated its agreement with this result. Its properties make it eminently suitable as a non-sweet replacement for the bulk and texture normally provided by sucrose, and its safety has been carefully established, leading to the recent food additive approval by the Food and Drug Administration.

ACKNOWLEDGMENTS

There is never enough space to acknowledge all contributors adequately. However, special recognition is due Hans Rennhard, the inventor, who also did the structure and much of the analytical work, Andy Torres for his expert guidance in the applications area, and John Beereboom, who kept us all on the right track.

REFERENCES

Actuaries, Society of, Chicago (1959). "Build and Blood Pressure Study."

Dwyer, J. T. (1972). *Postgrad. Med.* *51* (5), 91.

Figdor, S. K., and Rennhard, H. H. (1981). *J. Agric. Food Chem.* *29*, 1181-9.

Krall, L. P., and Joslin, A. P. (1971). In "Joslin's Diabetes Mellitus" (Marble, White, Brady, Hall, eds.), 11th edition. Lea and Febiger, Philadelphia.

MacDonald, I. (1971). In "Sugar," chapter 17. Butterworth's, London.

Mayer, J. (1969). "Overweight." Prentice-Hall, Englewood Cliffs, New Jersey.

Mayer, J. (1972). *Postgrad. Med.* *51* (5), 66.

McGandy, R. B., *et al.* (1967). *New Engl. J. Med.* *277* (4-5), 186-242.

Nelson, R. A., *et al.* (1973). *JAMA* *223* (6), 627.

Rennhard, H. H. (1973). U. S. Patent 3,766,165.

Sassoon, H. F. (1973). *Am. J. Clin. Nutr.* *26* (7), 776.

Yudkin, J. (1972). *Nature* *239*, September 22.

THE FUNCTIONAL APPLICATIONS OF POLYDEXTROSE

R. E. Smiles

Pfizer Inc.
Groton, Connecticut

I. INTRODUCTION

This chapter will cover the properties, functionality, applications and labeling requirements of Pfizer's fascinating new water soluble, reduced calorie bulking agent - polydextrose.

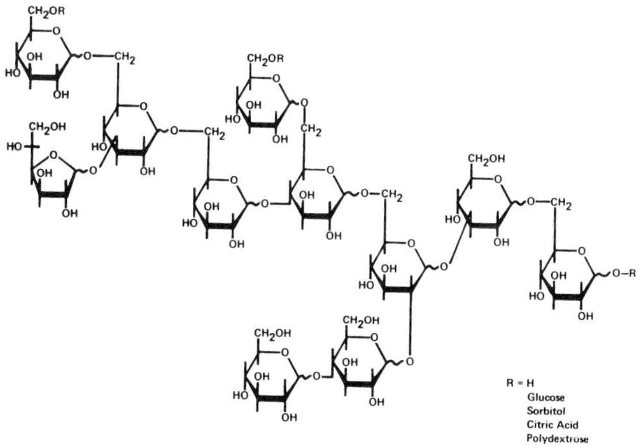

FIGURE 1. Polydextrose Chemical Bonding

ISBN 0-12-169080-6

Since we are creatures of habit, we like to eat those foods that give us pleasure such as cakes, candies, cookies, ice cream and puddings but we are also aware of the consequences that may follow. Excess body weight is a problem for many people. Since controlling weight to a desired level is a difficult task, there is a need for reduced calorie foods.

The beverage industry has partially met this need with diet soft drinks and light beer. When it comes to other pleasurable products - reduced calorie desserts and confections - the desires of the diet conscious consumer have not been satisfied.

In solid foods, you can replace the sweetness of sugar with an artificial sweetener, but how do you replace the bulk of the removed sugar? Until now the food additives geared to achieve caloric reduction in solid foods have been inadequate or non-existent. In some cases, a reduced calorie product such as hard candy could not be achieved and in other cases, consumers often perceived that caloric reduction resulted in inferior quality.

This is why polydextrose was developed: to satisfy the need for a water soluble reduced calorie bulking agent which could substitute for sugar and, in some cases, fat. Polydextrose can preserve a food's desirable attributes while removing some of the empty calories of sugar and the potential hazards of excess dietary fat.

II. PRODUCT DESCRIPTION

Polydextrose having 1 calorie per gram has exceptional utility in the preparation of reduced calorie and low calorie foods for weight control. When conventional foods are modified for dietary purposes, texture and body sometimes are adversely affected. Polydextrose provides not only the bulk but also gives the appropriate texture and mouthfeel qualities normally provided by sugar and fat.

Polydextrose, while having many of the technological properties of sugar, does not have sweetness. In some applications, this lack of sweetness is advantageous because the conventional sugar-based food is too sweet for some tastes. The combination of polydextrose and a sweetener permits the sweetness level to be adjusted over a wide range. Products having one-third fewer calories often can be prepared by replacing a portion of the sugar normally present in foods with polydextrose. In foods where it is desirable to restrict both calories and sugar, an artificial sweetener can be used. When polydextrose is used with an artificial sweetener, calories often can be reduced by 50% or more.

Polydextrose is available in two forms: polydextrose (powder); polydextrose type N (70% solution).

Polydextrose is an amorphous, white to light tan, water soluble powder. It contains residual acidity providing a pH range of 2.5 to 3.5. It forms a clear melt about 130°C in a manner similar to sucrose.

Polydextrose type N is a clear, colorless to yellow, 70% solution made by partially neutralizing polydextrose with potassium hydroxide. It has a pH range of 5 to 6. In many foods, either form may be used.

Along with the polymer itself, there are small amounts of the residual starting materials, namely glucose, sorbitol and citric acid. In addition small amounts of levoglucosan and hydroxymethyl furfural are produced as a result of caramelization of glucose. (Table I)

TABLE I. Polydextrose[a] Composition

	(%)
Polymer	> 90
Glucose	< 4
Sorbitol	< 2
Levoglucosan	< 4
Water	< 4
Sulfated Ash	< 0.3

[a]*Monomer and polymer (anhydrous, ash free basis).*

III. PROPERTIES

A. *Solubility*

Aqueous solutions can be prepared from polydextrose powder. The rate of solution of the powder depends on the speed and shear of the mixing equipment and the manner in which the powder is added to the water. In preparing concentrated solutions, difficulties in dissolving polydextrose in water can be overcome by slow addition and efficient mechanical dispersion of the polydextrose in hot water. The addition of a second soluble material as a dispersing agent to the dry powder will also facilitate solution.

B. Viscosity

The viscosity of polydextrose solutions is somewhat greater than that of sugar solutions of equal concentrations. Figure 2 shows the viscosity/temperature relationship for 70% solutions of polydextrose and sucrose. The viscosity/concentration relationship of both compounds is shown in Figure 3.

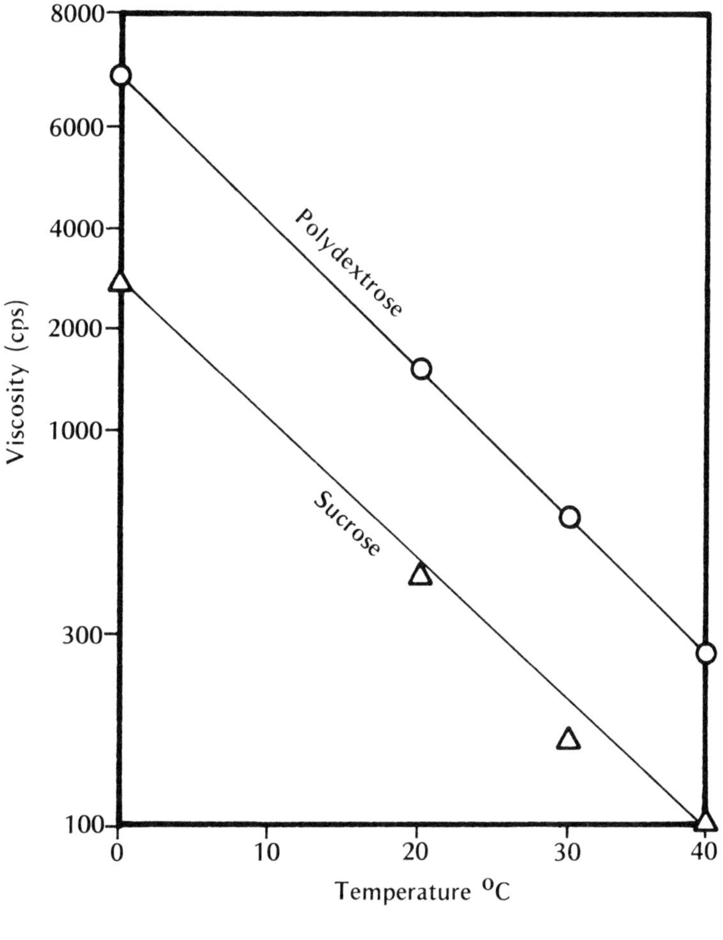

FIGURE 2. Temperature/viscosity relationship of a 70% solution of Polydextrose and a 70% solution of sucrose.

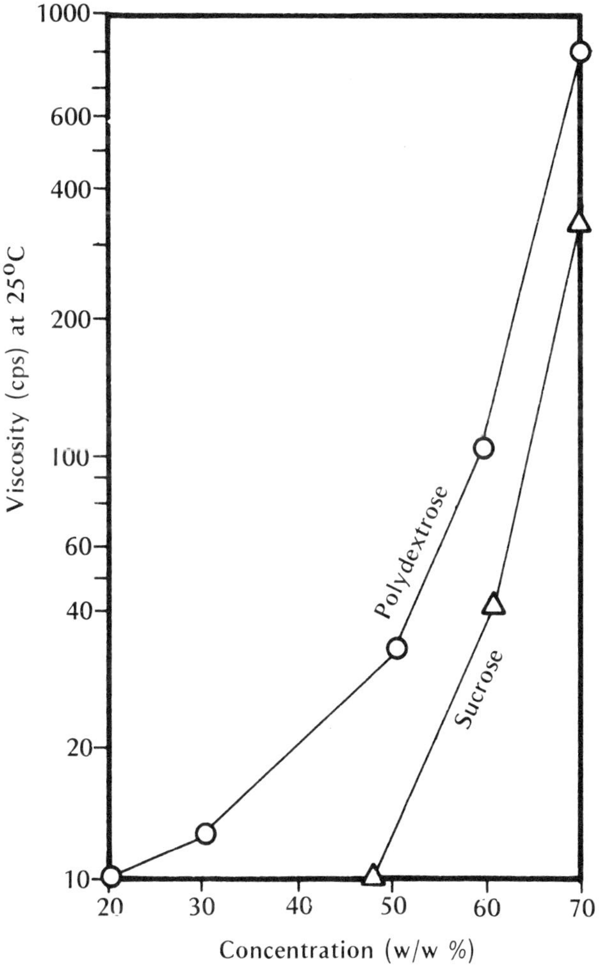

FIGURE 3. Concentration/viscosity relationship of polydextrose solution and a sucrose solution.

C. Humectancy

Polydextrose can function as a humectant to prevent or slow undesirable changes (either an increase or a decrease) in the moisture content of food. Figure 4 shows how polydextrose powder absorbs moisture under controlled atmospheric conditions until an equilibrium, which is characteristic of the relative humidity, is reached. Figure 5 demonstrates how polydextrose type N (70% solution) loses water to attain the same equilibrium. In hard candy and baked goods, polydextrose can help control the rate of moisture gain or loss during prolonged periods and therefore can serve as an important additive in extending shelf life.

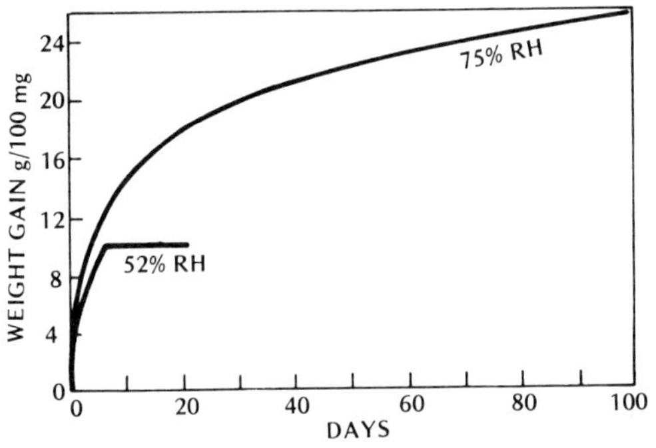

FIGURE 4. Hygroscopicity of polydextrose at 25°C and two different relative humidities.

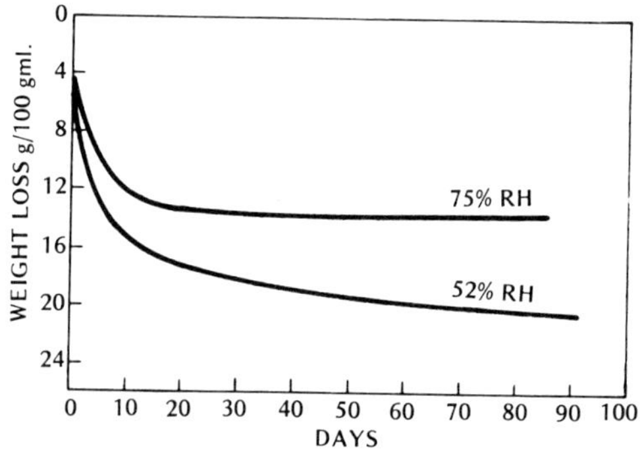

FIGURE 5. Loss of water from polydextrose type N solution at 25°C and two different relative humidities.

D. Stability

Polydextrose and polydextrose type N both had excellent storage stability over a 90 day investigative period at 25°, 45° and 60°C. Generally, minor hydrolytic changes were observed but these were slow to occur and are considered not to be of practical importance under expected conditions of use. The only significant change was observed at elevated temperatures where the polydextrose type N exhibited a darkening in color. The higher the temperature, the greater the rate of change. For this reason storage of the type N in a cool place is recommended.

E. Sweetness Value

Polydextrose has no sweetness. Consequently it can provide improved mouthfeel and textural qualities in foods without adding excess sweetness.

F. Caloric Value

Polydextrose is only partially metabolized to the extent that its caloric contribution is 1 calorie/gram compared to 9 calories/gram for fat and 4 calories/gram for sugar.

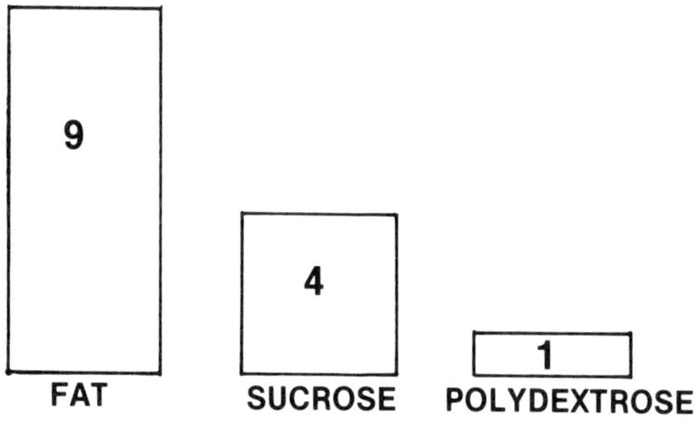

FIGURE 6. Caloric utilization values. (calories/gram)

IV. FUNCTIONALITY

Polydextrose can function as a bulking agent, formulation aid, humectant and/or texturizer as defined in federal regulation 21 CFR 172.841. It can be used in certain food products (which are discussed in the application section of this chapter) in accordance with good manufacturing practices when standards of identity do not preclude its use. No maximum use levels have been set by the Food and Drug Administration other than the amount required to achieve the intended functional effect in the food products.

A. Bulking Agent

As a bulking agent, polydextrose contributes solids to maintain palatability and textural qualities without sweetness. Its one calorie per gram aids in the development of reduced and low calorie foods.

B. Formulation Aid

As a formulation aid, polydextrose improves the flowability of other ingredients in a dry mix. In preparing a finished product, uniform dispersion and the prevention of severe lumping of other solid ingredients are benefited with the addition of polydextrose.

C. Humectant

The humectant aspect of polydextrose controls moisture and can thereby provide tenderness and softness in such products as baked goods and candy.

D. Texturizer

As a texturizer, polydextrose provides improved mouthfeel and viscosity qualities. It can be used to control, modify, and improve texture in formulated food products.

V. APPLICATIONS

The efficacy of polydextrose as a partial replacement for sugars and/or fats resulting in a significant reduction in utilizable calories has been demonstrated in a variety of foods.

At the present time, polydextrose has been accepted as a new food additive (21 CFR 172.841) for use in the following eight food categories:

1. Baked goods and baking mixes (restricted to fruit, custard and pudding filled pies, cakes, cookies and similar baked products).
2. Chewing gum.
3. Confections and frostings.
4. Dressings for salads.
5. Frozen dairy desserts and mixes.
6. Gelatins, puddings and fillings.
7. Hard candy.
8. Soft candy.

By federal regulation (21 CFR 105.66) a food can be called "low calorie" only if it has no more than 40 calories per serving and no more than 0.4 calories per gram. A food may be labeled "reduced calorie" on the basis of an alteration or fabrication of special dietary significance that results in a calorie reduction of at least one-third.

The following formulations have been developed to serve as demonstration vehicles in the application of polydextrose. These represent a one-third calorie reduction over conventional sugar and/or fat controls. The appearance, taste and textural properties were evaluated by sensory panel hedonic rating techniques. In general, the polydextrose products received acceptable ratings when compared to their conventional counterparts.

A. Baked Goods and Baking Mixes

In the baked goods and baking mixes category, the use of polydextrose is restricted to fruit, custard and pudding filled pies - cakes - cookies and similar baked products.

In cakes, polydextrose functions as a tenderizing agent and is used to replace sugar and fat. No significant changes in volume, structure, tenderness or flavor were observed. Polydextrose powder with a pH range of 2.5 to 3.5 was used. Sufficient sodium bicarbonate was added to neutralize the acidity. The addition of a gum such as xanthan was included to improve the cake batter stability. Additional modification of the texture can be achieved by using a combination of polydextrose and fiber type ingredients such as microcrystalline cellulose.

Chocolate Cake

	Ingredients	*Percent*
Part A	*Sugar, granulated*	*16.50*
	Polydextrose	*14.50*
	Hydrogenated vegetable shortening (Crisco)	*4.50*
	Mono- and diglycerides (Atmul 500)	*0.60*
	Sodium bicarbonate	*0.40*
	Artificial chocolate flavor	*0.30*
	Artificial vanilla flavor	*0.05*
Part B	*Cake flour*	*15.51*
	Crystalline cellulose (Solka-Floc BW-200)	*6.50*
	Nonfat milk solids	*1.60*
	Salt	*0.30*
	Baking powder (Davis)	*0.34*
	Sodium bicarbonate	*0.58*
	Dicalcium phosphate	*0.38*
	Sodium aluminum phosphate (BL-60)	*0.35*
	Cocoa, dutched (10-12% fat)	*3.50*
	Xanthan gum (Keltrol)	*0.07*
Part C	*Whole eggs*	*17.82*
	Water	*8.10*
Part D	*Water*	*8.10*

The chocolate cake is prepared by creaming the ingredients in part A. The ingredients in part B are then slowly added and mixed. The whole eggs and water in part C are then added and mixed until smooth. The remaining water from part D is added and mixed until smooth. Sixteen ounces of the mix is placed in an eight inch cake pan (greased or paper lined) and baked at 370°F for 30 minutes.

Among the acceptable polydextrose baked goods developed were chocolate, yellow, pound and sponge cakes, brownies, chocolate and butter cookies.

B. Frostings

In a chocolate frosting, polydextrose was used as a partial replacement for sugar and fat. The weak acidity of polydextrose was overcome by the use of a food approved neutralizing agent such as sodium bicarbonate. Because of polydextrose's lack of crystallinity, a change in texture toward a taffy type was noted.

Chocolate Frosting

	Ingredients	*Percent*
Part A	*Butter*	5.96
	Mono- and diglycerides (Atmul 500)	1.60
	Polysorbate 60 (Tween 60)	0.26
Part B	*Powdered sugar 10X*	41.73
	Sodium bicarbonate	0.47
	Salt	0.23
Part C	*Polydextrose*	25.73
	Crystalline cellulose (Solka-Floc BW-300)	6.72
	Cocoa, dutched (10-12% fat)	3.01
Part D	*FD&C Color Solution*[a]	0.41
	Vanilla extract	0.63
	Water	13.25

[a]*FD&C Red 40 - 4.57% solids*
FD&C Yellow 5 - 4.44%
FD&C Blue 1 - 0.99%
q.s. to 100 ml with water

Melt part A and then add to part B and cream. Add part C and blend. While mixing, slowly add part D; scrape down and whip until smooth.

C. Chewing Gum

Acceptable chewing gum can be prepared by using a combination of polydextrose and sorbitol or sugar. The sweetness quality can be enhanced with an artificial sweetener such as saccharin or aspartame. Greater elasticity and shelf life were noted as positive attributes.

D. Frozen Dairy Desserts

Good quality non-stardandized frozen desserts can be prepared using polydextrose type N as a partial replacement for sugar and butterfat. The product prepared in a commercial batch-type freezer closely resembled the creaminess, smoothness, mouthfeel and flavor qualities of a standard ice cream. The polydextrose served the same function as sugar in depressing the freezing point. The melting rate and overrun were comparable to a sucrose standard.

Vanilla Frozen Dessert

Ingredients	*Percent*
Heavy cream, 35% butterfat	*14.50*
Polydextrose Type N (70% solution)	*13.91*
Water	*52.90*
Sugar, granulated	*9.74*
Nonfat milk solids	*7.32*
Glycerol monostearate (Drewmulse 700E)	*0.30*
Gelatin 120 Bloom	*0.33*
Vanilla extract	*1.00*

The frozen dessert is prepared by combining the heavy cream, polydextrose, and water. The dry ingredients, sugar, milk solids, glycerol monostearate and gelatin are then added and the mixture heated to 45°C, while stirring to dissolve the solid ingredients. The temperature is raised to 71°C for 30 minutes to pasteurize. The mixture is homogenized at 2500 psi (first stage) and at 500 psi (second stage. The homogenized mixture is quickly cooled to 4°C and aged 18-24 hours at 5°C. The vanilla extract is added, and the mixture frozen and stored according to conventional practice.

E. Dressings for Salads

A low calorie liquid nonstandardized French-type dressing was prepared, substituting polydextrose for thc vegetable oil. Uniform viscosity and creamier mouthfeel were noted.

F. Puddings

In instant puddings, the addition of polydextrose produced a creamier texture, and a more uniform dispersion of cocoa and starch.

In artificially sweetened instant puddings, polydextrose acts as a formulation aid, and prevents the severe lumping, nonuniform dispersion of the solids and the poor set encountered in products without polydextrose.

Chocolate Instant Pudding

Ingredients	*Percent*
Fructose	*33.40*
Polydextrose	*20.23*
Food starch modified H50B (National Starch)	*23.70*
Cocoa, dutched (10-12% fat)	*17.20*
Sodium phosphate, dibasic	*3.20*
Sodium pyrophosphate	*1.72*
Salt	*0.43*
Vanillin	*0.12*

A uniform blend of the ingredients is prepared. 93 grams of the pudding mix is added to 2 cups of cold skim milk, in a mixing bowl. Mix at low speed with an electric mixer, until well blended and pour into serving bowls.

To provide additional sweetness, aspartame can also be considered in this application.

G. Hard Candy

Hard candy was prepared by replacing 50% of the sugar normally present with polydextrose. The candy had reduced sweetness, but the texture, flavor and appearance was the same as an all-sugar candy. Since polydextrose is acidic, small amounts of sodium citrate must be added to reduce the tendency for sugar inversion during the cooking process.

Hard Candy

Ingredients	*Percent*
Polydextrose	*38.65*
Sugar	*38.65*
Sodium citrate (dihydrate)	*0.45*
Water	*20.50*
Citric acid (anhydrous)	*1.50*
Lemon flavor	*0.15*
FD&C Yellow 5, 5% solution	*0.10*

The hard candy is manufactured by dissolving the polydextrose, sugar, and sodium citrate in the water with heat and agitation, being careful to dissolve all the crystals from the side of the vessel. Cook to 155°-160°C and discontinue heating. When the candy mass has cooled to 130°-140°C, add acid, flavor, and color with agitation. Deposit into molds or use conventional stamping equipment to form the finished piece.

In hard candy, excessive moisture gain is undesirable and should be prevented. Moisture gain under conditions of controlled humidity indicated that sucrose candies absorbed water substantially faster, resulting in greater stickiness than the polydextrose counterpart. Polydextrose significantly slowed the deterioration of the candies.

Peanut brittle prepared with a polydextrose-sugar combination was equivalent in flavor and texture to a standard product containing sugar and corn syrup. This product was buffered with sodium citrate to pH 4.3 to 4.5 to prevent sugar inversion. Polydextrose produces a significantly lighter colored and less tacky product with one-third fewer calories than a typical sugar product.

Peanut Brittle

Ingredients	*Percent*
Polydextrose	*33.17*
Water	*15.17*
Sodium citrate (dihydrate)	*0.34*
Sugar	*35.91*
Full fat raw peanuts	*4.28*
Defatted peanuts	*8.88*
Butter	*1.11*
Sodium bicarbonate	*0.94*
Salt	*0.20*

This peanut brittle is made by dissolving the polydextrose, sugar and sodium citrate in the water and heating to 135°C while stirring. Raw peanuts, defatted peanuts and butter are added in sequence while the cooking proceeds to 145°C, 150°C and 155°C after each addition. Heating is discontinued and the sodium bicarbonate and salt are added when the mass cools to 150°C. Stir until well mixed and the mass puffs up. Restir to puff the mass again and pour onto an oiled slab. Quickly spread the batch to form a thin product and break it into pieces when cool.

H. Soft Candy

Polydextrose helps retain softness of soft candy because of its humectant properties. This is in addition to providing bulk and reduced calories. Complete replacement of corn syrup can be achieved in most confectionery products with no loss in eating qualities. Formulations for gum drops and caramel having a one-third calorie reduction were developed.

Gum Drops

	Ingredients	Percent
Part A	Starch (Eclipse 'F' Staley)	8.30
	Water	18.70
Part B	Polydextrose	23.50
	Sugar	19.50
	Sodium citrate (dihydrate)	0.50
	Water	29.46
Part C	Strawberry flavor	0.02
	FD&C Red 40, 10% solution	0.02

Prepare a suspension of the starch by adding starch to cold water (part A). Combine the ingredients in part B and bring this solution to a boil. Add the prepared starch suspension (part A) gradually to part B. Continue boiling slowly with stirring until the solids level reaches 76% by refractometer. Add the flavor and color (part C) mixing well. Deposit into starch molds and store at about 100°F until 80% solids are attained. This takes about 30 hours. Lightly dust the gum drops with granulated sugar.

Hear again, the sodium citrate functions as a buffer to prevent sugar inversion.

Caramel

Ingredients	Percent
Polydextrose type N	32.46
Sugar	25.13
Medium cream (25% fat)	12.70
Glycerol monostearate (Drewmulse 200K)	1.00
Water	5.20
Evaporated milk	23.46
Vanillin	0.05

The caramel is made by combining the polydextrose solution, sugar, cream, glycerol monostearate, and water with agitation and heating until the mixture comes to a boil. Add the evaporated milk and continue stirring and heating to a temperature of 116°C. Discontinue the heat, thoroughly stir in the vanillin and pour the mass onto a lightly oiled slab. When cooled to room temperature, the product is cut to desired size and wrapped.

VI. LABELING AND NUTRITONAL INFORMATION

Food products containing polydextrose designed for special dietary use, such as reduced or low calorie foods, must be labeled in accordance with 21 CFR Part 105.

The ingredient statement of foods containing polydextrose (either the powder or the 70% solution partially neutralized with potassium hydroxide) should include the officially recognized name of the ingredient "polydextrose. The potassium hydroxide used in preparing the 70% solution is considered an incidental additive and need not be labeled when the potassium content per serving is less than 50 mgs. Foods containing in excess of the 50 mgs potassium per serving must also declare potassium hydroxide in the ingredient statement. A clarifying statement may also follow such as "for neutralization" or "to adjust pH."

FDA recognizes that polydextrose contains only one calorie per gram and does not object to the use of this value for purposes of determining the caloric content of foods formulated with polydextrose.

As expected for a bulking agent, a laxative effect can occur at high levels of ingestion. In a study with adults, the mean laxative threshold was 90 grams per day for polydextrose compared to 70 grams per day for sorbitol. Polydextrose, therefore, is well tolerated in amounts likely to be consumed in foods for which it is approved and is expected to cause fewer problems than currently used polyols such as sorbitol and mannitol.

When a single serving of a food contains more than 15 grams of polydextrose, the label of that food must bear the statement: "Sensitive individuals may experience a laxative effect from excessive consumption of this product."

VII. SUMMARY

In summary, polydextrose provides the food technologist with an effective new tool to design products for today's more attractive lifestyle. It is a water soluble, low calorie bulking agent which can substitute for sugar and fat. It has no sweetness and contributes only 1 calorie per gram.

Polydextrose can function as a bulking agent, formulation aid, humectant and texturizer and is presently accepted for use in eight food categories.

With polydextrose, a variety of reduced calorie food products can be created which simulate the higher calorie conventional counterparts in appearance, palatability, texture and mouthfeel.

INGREDIENTS USED IN EXTRUSION-COOKED FORMULATED FOODS SHOULD BE SELECTED FOR THEIR ORGANOLEPTIC, FUNCTIONAL, AND NUTRITIONAL PROPERTIES

Oak B. Smith

WENGER INTERNATIONAL, INC.
Kansas City, Missouri

Thermodynamic evaluation of available methods of thermal processing shows that High Temperature/Short Time (HT/ST) extrusion cookers are more efficient thermodynamically than older thermal processes. Extrusion cooked foods are being developed by a technological orientation of the principles of thermodynamics with selection of extruder components and processing controls, and ingredients to produce desired functional and organoleptic characteristics in fabricated foods. Nutritional considerations weigh as heavily today on food scientists as does the selection of ingredients to produce desired organoleptic or functional characteristics in the food to be fabricated.

These considerations combine today to produce a whole bevy of fabricated foods -- breakfast cereals, snacks, textured soy proteins, instantized hot breakfast gruels, breadings, infant foods, quick cooking pastas, confectionaries, croutons, full fat soy proteins, and precooked, protein fortified, vitamin and mineral enriched structured foods in the form of beverage powders, finger foods, dry soup mixes, pastas, or as foods packaged for freezing, or as meals in a pouch.

The selection of ingredients to produce desired organoleptic characteristics in the fabricated foods is one of the major tools of the food scientist. The selection of pH modifiers, gums, emulsifiers, thickeners, and surface active ingredients, can be as important in food technology as the selection of flours, fiber, or oils to produce the desired density, shape, or mouthfeel in fabricated foods.

ISBN 0-12-169080-6

THERMODYNAMIC EFFICIENCY

The thermodynamic efficiency of HT/ST extrusion cookers will compare very favorably with older methods of thermal processing -- pressure cooking, boiling, baking, spray drying, canning, drum drying, etc.

A group whose discipline is rooted in thermodynamic evaluations (Berry 1972, Slesser 1973, Hawthorn 1973 and others) have enunciated a theory which I believe to be valid, constructive, and germane to the needs of a world which finds itself in an energy crisis. In effect, this theory says that the world may no longer be able to afford some of its food habits -- that we need now to evaluate every aspect of food production from the standpoint of total energy inputs. They refer not only to the energy inputs of a food processing system, but also to the energy "subsidy" which goes into every step of production of grains, legumes, or oilseeds. They refer to the energy "subsidy" in producing fertilizer or pesticides, in the preparation of the soil, and in seeding, cultivating, irrigating, harvesting, drying, milling, extracting, food processing, particle sizing, pasteurizing, drying or canning, packaging, refrigerating, freighting, distributing, and in the energy subsidy used in the cooking of foods at the point of consumption, and presumably including the disposition of effluents and ecological hazards at each point in the food production and food preparation chain. The same type of evaluation needs to be extended to all aspects of meat production. When one looks at the consumption of energy in this light, it seems essential that every step in the food production chain must be efficient in terms of energy consumed, be it horsepower, manpower, or thermal energy.

When we examine food costs in the light of total energy subsidy plus the total labor costs, and relate those costs to 100 grams of utilizable protein, or to 100 calories of metabolizable energy, I believe we may find that many of our food patterns should indeed be changed. It takes ten kilograms of grain/protein feed to produce one kilogram of gain on beef cattle. On top of that, a cow has to be fed for one year to produce a calf for fattening in this energy intensive beef production chain.

Slesser (1) points out that "energy utilization to power the human race is just as subject to the laws of thermodynamics as the energy utilization to power an internal combustion engine." Berry (2) has shown that "thermodynamic analysis is a realistic way to assess the ultimate value of an economic process because it does so in terms of the only commodity in the world in ultimate limitation -- thermodynamic potential". "If the economists in the marketplace were to determine their shortages by looking further and further into the future, these estimates would come closer and closer to the estimates made by their colleagues, the thermodynamicists".

HT/ST extrusion cookers are thermodynamically efficient converters of electrical energy to thermal energy. They efficiently utilize modest levels of steam, but most of the added moisture is applied as water. They require less labor per ton of production than any other cooking method. They have the ability to entrain moisture and heat; make doughs; gelatinize starches; denature and debitter proteins; control heat labile growth inhibitors; pasteurize and histologically restructure process materials; inactivate enzymes, and retexture, shape, expand, size and partially dry and cool the extrusion cooked extrudate. They bond all ingredients together uniformly and irreversibly.

It may be realistic to suggest that a process which improves the utilization and acceptability of cereal and oilseed proteins will meet Berry's criteria for thermodynamic potential when it does so with greatly lower total energy requirements per ton of production within this thermodynamic concept than any other industrial cooking method.

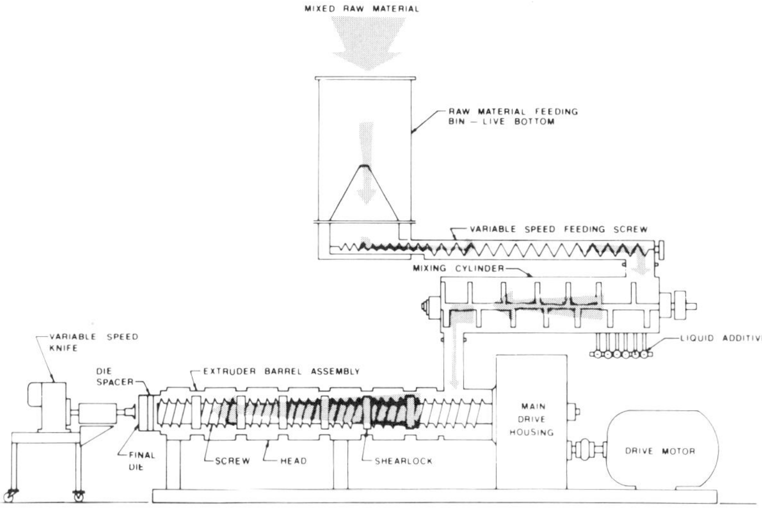

FIGURE 1. Cutaway view of HT/ST extrusion cooker.

It is suggested further that there is a considerable bonus in terms of thermodynamic potential when one compares the thermodynamic costs of producing extrusion cooked textured soy meat extenders against the thermodynamic costs of producing beef, or pork, or most crustaceans.

FABRICATED FOODS

Scrimshaw (3) points out that protein supplied in the human diet from cereals exceeds that from all other sources, and for a majority of the world's population cereal staples furnish two thirds or more of their total protein intake.

It seems obvious that many new fabricated foods will be based on cereal proteins, oilseed proteins and/or pulse proteins, and that meat and dairy proteins may have a lesser part to play in the development of fabricated foods.

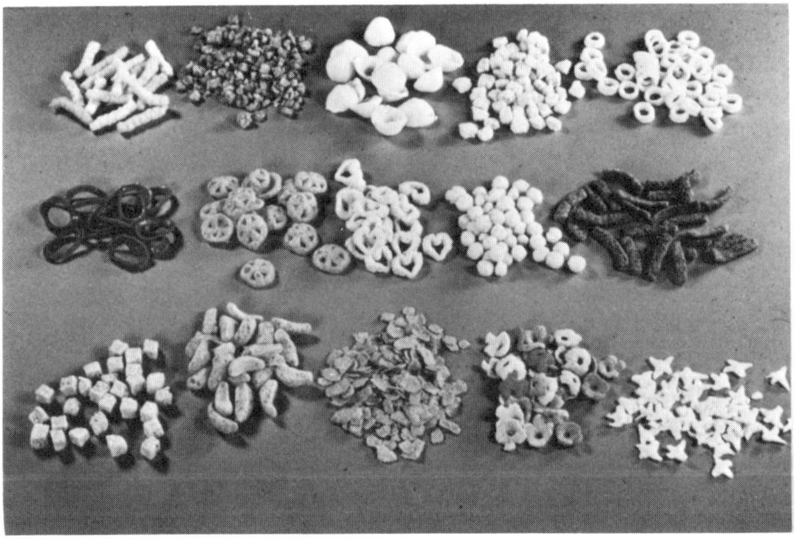

FIGURE 2. Extrusion cooked fabricated foods.

FIGURE 3. Extrusion cooked breakfast cereals produced abroad.

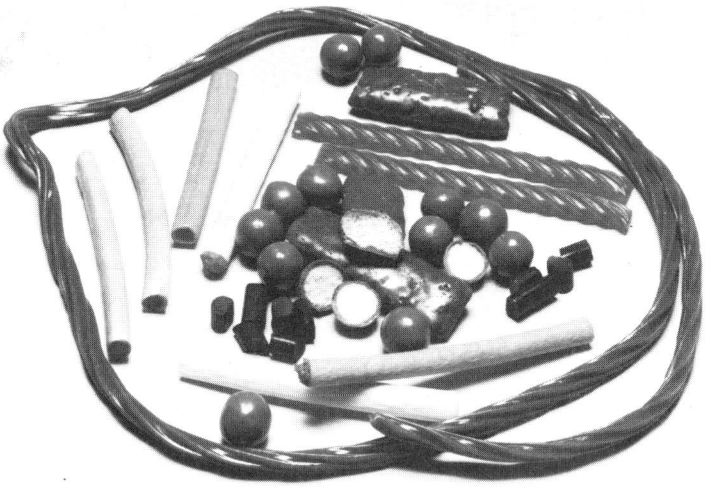

FIGURE 4. Extrusion cooked confectionery products.

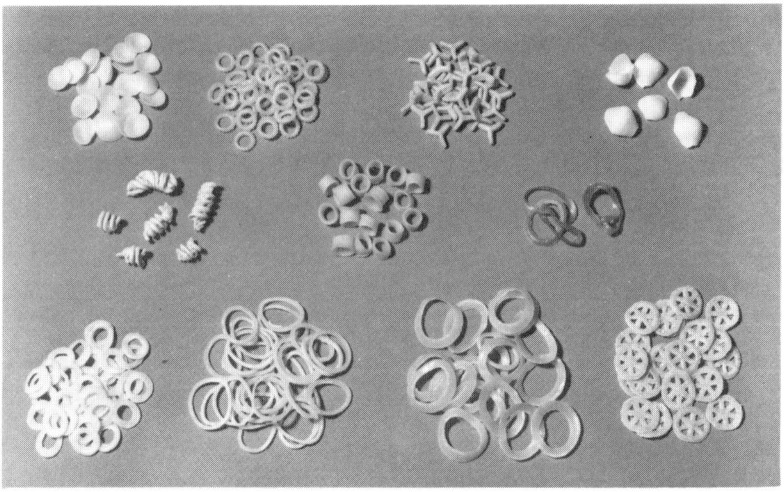

FIGURE 5. Double extruded third generation snacks before french frying.

FIGURE 6. Breakfast cereal flakes produced from twice extruded beads which are rolled to flakes on roller mills.

FIGURE 7. Textured soy meat extenders produced from defatted soy flour or soy concentrates.

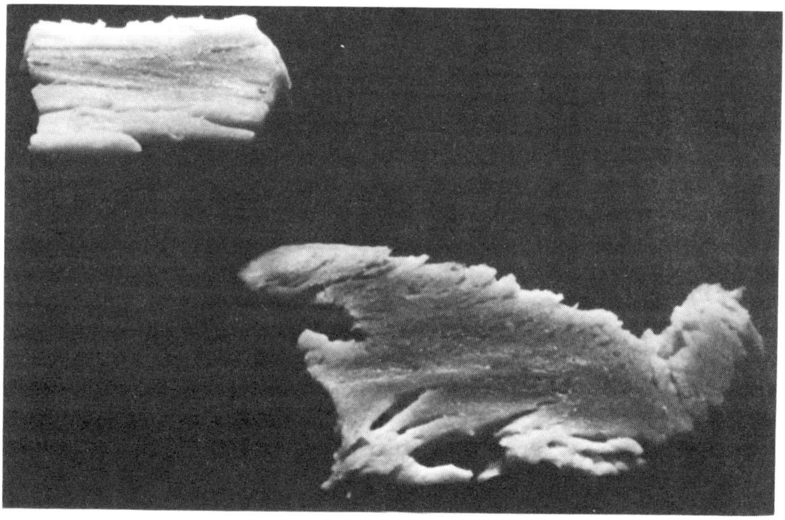

FIGURE 8. Uni-Tex meat analog produced from defatted soy flour or concentrates, and vital wheat gluten. Additionally, it would be possible to use rapeseed concentrates produced from the double zero varieties of rapeseed when such usage is permitted.

FIGURE 9. Quick cooking pasta products produced from wheat flour or rice flour.

FIGURE 10. Instantized legume powders produced for dry soup mixes.

FIGURE 11. Extrusion cooked wafers, cookies, and crisp breads.

FIGURE 12. Instantized hot breakfast gruels produced from oat flour, rice flour, or wheat flour, with or without wheat bran.

FIGURE 13. Cold breakfast cereals produced from second clears of wheat flour plus wheat bran in the form of bran flakes, bran pillows or bran wheels.

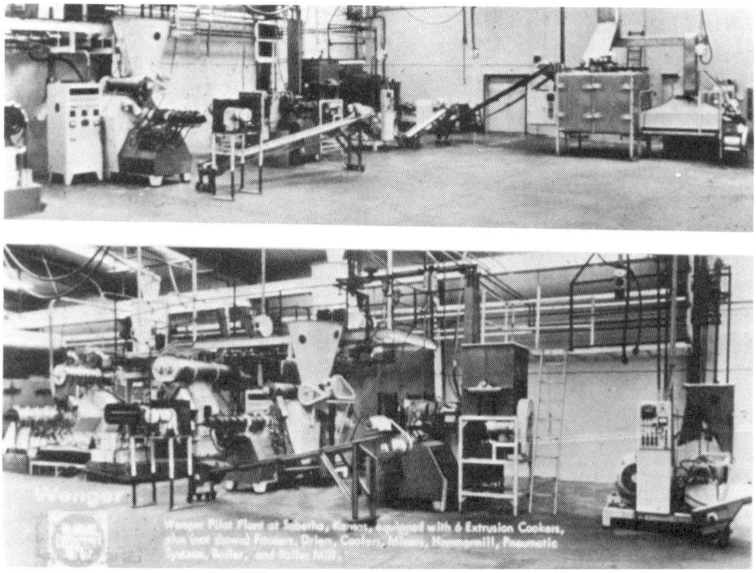

FIGURE 14. Wenger pilot plant is equipped with eight extrusion cookers, two dryers, water jacketed roller mill, batch mixer, impact mills, pneumatic conveyors, and a laboratory fitted with moisture testers, viscosity measurement devices, gas chromatography, pH measurements, etc.

Wenger's pilot plant operates about 300 days per year developing new fabricated foods, or improving on existing foods.

FIGURE 15. Automatic processing controls with extrusion cooker.

Automatic control system measures temperatures at several points in the system, controlling preselected temperatures automatically by control of circulation of cooling water or steam through jackets around segmented extruder heads. The automatic control unit can measure (and record if desired) temperatures and pressures within an extruder, and control application of water, steam, feeder RPM and extruder RPM. This unit is fitted with push-button controls and indicator lights for all motors used in the system. Tachometers are installed in the control panel to measure RPM when variable speed drive motors are used on the feeder assembly. Ammeters are also installed for measuring electrical usage for the larger motors in the system. This photograph also pictures a secondary cooling and forming extruder, about which more will be said later in this chapter.

CONTROL OF PROCESSING VARIABLES

One of the most valuable tools for the food scientist working with extrusion cooking is the knowledge that HT/ST extrusion cookers are designed to provide flexibility in several major mechanical ways:

1. Control of process temperatures. Temperatures are controlled to an important extent by selection of extruder screw and sleeve design used in each extruder head. Johnston (4) has categorized the design of extruder screws into three distinct groups -- low shear, medium shear and high shear extruders.

	LOW SHEAR	MEDIUM SHEAR	HIGH SHEAR
PRODUCT MOISTURE %	25-75	15-30	5-8
PRODUCT DENSITY (GRAMS/LITER)	320-800	160-510	32-200
MAXIMUM BARREL TEMPERATURE (°C)	20-65	55-145	110-180
MAX. BARREL PRESSURE (kg/cm^2) (kPa)	6-63 588-6 178	21-42 2 059-4 119	42-84 4 119-8 238
$\frac{Ds}{h}$ (SCREW DIAMETER / CHANNEL DEPTH)	3-5.3	5.0-8.5	8.0-18.0
PARALLEL FLOW CHANNELS (n)	1	2	2 or 3
SCREW SPEED (RPM)	LESS THAN 100	GREATER THAN 200	GREATER THAN 200
ENERGY CONVERSION (kW/kg)	0.01-0.04	0.02-0.08	0.10-0.16
TYPICAL PRODUCTS	PASTA PRODUCTS THIRD GENERATION SNACKS MEAT PRODUCTS GUMS	TEXTURED SOY BREADING EXPANDED PET FOOD SEMI MOIST PET FOOD	SNACKS BREAKFAST CEREALS BREADING (CROUTONS) THIN BOILING STARCH

FIGURE 16. Single screw extruder classification illustrating general characteristics of foods dependent on shear strength of individual products.

General characteristics of foods dependent on shear strength of individual ingredients are noted by Johnston for each major category of extruder configuration.

2. Control of moisture application. Moisture is controlled by meters measuring water flow or steam flow into the high speed mixer and/or into the extruder assembly. Uniformity of moisture application is critically important in extrusion cooking. For this objective, the system is designed to provide uniformity of water pressure and steam pressure. Meters for water flow and steam flow are important. Of equal importance is an uniform adjustable rate of flow of mixed dry ingredients to the high speed mixer.

3. Control of dwell time. It may be desirable when switching production from one product to another to vary the dwell time within the extruder. This can be accomplished by change in extruder RPM, or by exchanging designs of extruder screw or shearlock dies. It is also possible to add extruder heads when more dwell time is needed, or to reduce the number of extruder heads when less dwell time is needed to produce a new product.

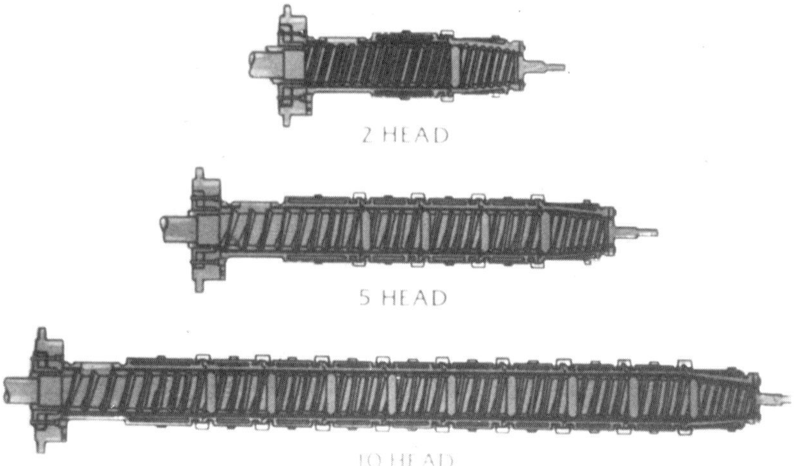

FIGURE 17. Illustrates lengthening or shortening of extruder dwell time.

Flexibility in mechanical design makes it possible to produce many products in HT/ST extrusion cooking. It is not yet possible to make bread by extrusion cooking, but HT/ST extrusion cookers are able to produce several varieties of cookies, crackers and crisp breads. We are working on production of breadsticks by extrusion cooking. To illustrate these potentials, Johnston (4) as a matter of convenience compares the conventional bread making process with extrusion cooking to graphically illustrate what is happening within the medium shear and high shear HT/ST extruder system.

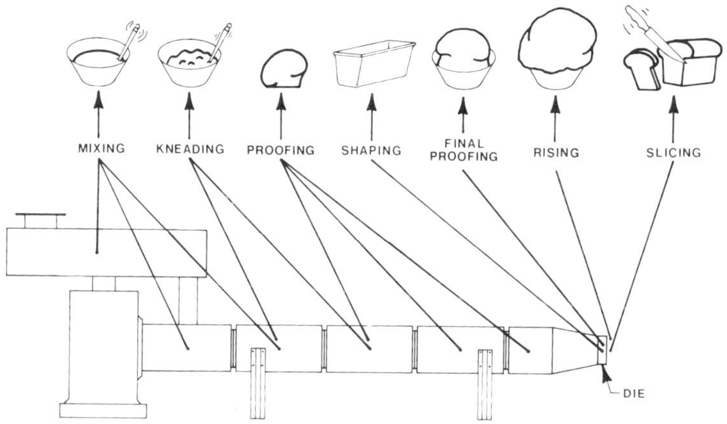

FIGURE 18. Conventional bread making compared to HT/ST extrusion cooking.

Breadings have been made traditionally in a manner similar to that shown in the conventional bread making process. It is now practical to make breadings most economically using the HT/ST extrusion cooking method as illustrated in Figure 19.

To make breadings on the HT/ST extrusion cooker as illustrated in Figure 19, it is additionally necessary to wet mill the breadings to the desired particle size, and thereafter to dry and cool them. When we compare the equipment costs and processing costs per ton in making breadings, crisp breads, breadsticks, wafers, etc., produced by conventional bread making methods compared to the equipment costs and the processing costs per ton of making these products by HT/ST extrusion cooking, I believe that Berry's thermodynamic analysis dramatically shows the great economic differences between the two processes.

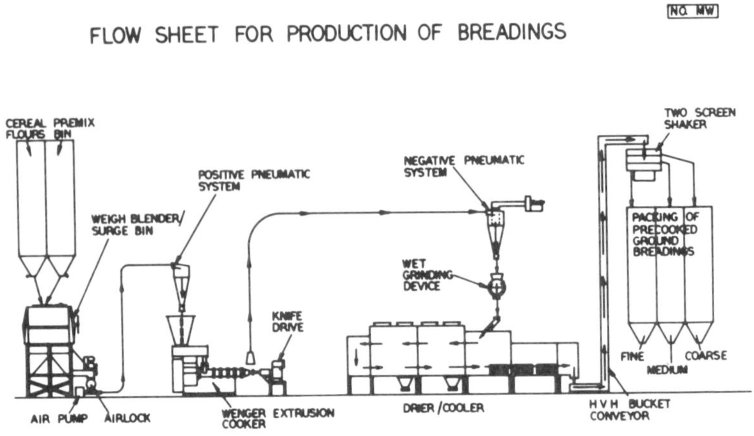

FIGURE 19. Flow sheet for the production of breadings.

DOUBLE EXTRUDED FOODS

There are several important categories of fabricated foods which are made by double extrusion -- by two extrusion units in Tamden. The first of these is a cooking extruder, in which starches are gelatinized, proteins are debittered and denatured, enzymes are inactivated, and the process materials are pasteurized, and bonded together irreversibly. The second extruder is a cooling and forming extruder, which gives desired textures, density, shape and structure to the finished product.

In this way, third generation snacks are densified and given desired shapes and structure. After being slowly dried and cooled, these products are french fried as onion rings, corkscrews, wheels, or chips for dips, etc.

FIGURE 20. Third generation snacks before and after french frying.

Another twice extruded product is breakfast cereal flakes -- corn flakes, wheat flakes, bran flakes, oat flakes or flaked mixtures of cereal flours, sometimes including defatted oilseed proteins. These products are cooked and expanded on the first extruder, and are densified on the second extruder to produce a small bead of dense and moist extrudate, which beads thereafter are rolled through into flakes using roller mills, and which flakes are subsequently dried and cooled.

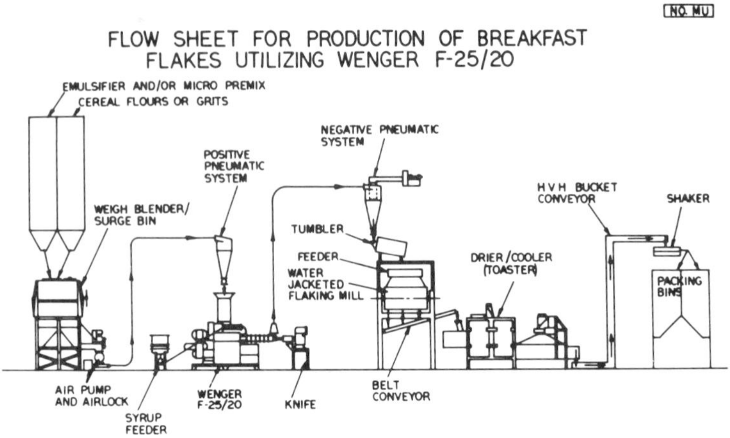

FIGURE 21. Flow sheet for extrusion cooking breakfast cereal flakes.

Another category of twice extruded product are textured soy meat analogs. These products should not be confused with textured soy meat extenders, which are also produced as textured soy proteins, but which are made on a single HT/ST extrusion cooker.

The textured soy meat analog is processed through a first extruder where the proteins are denatured and debittered, the growth inhibitors are destroyed, and the extrudate is fed into the second extruder. This second extruder cools and densifies the dough, and gives it a structure which has an uncanny resemblance to the mouthfeel and texture of meat - Smith (5) (see Figure 8).

FIGURE 22. Photograph of largest size of double extruder for production of meat analog "Uni-Tex".

SELECTION OF INGREDIENTS AND ADDITIVES

Selection of ingredients is one of the most important tools available to the food technologist to give desired functional characteristics and organoleptic properties to an extrusion cooked food.

Whenever any cereal and/or protein food is extrusion cooked, the food is texturally and histologically restructured. During the HT/ST extrusion cooking process the moistened granular or floury materials are converted into a dough. In the process, gelatinization of the starchy components occurs, resulting in great uptake of moisture and substantial increases in dough viscosity. During propulsion through the extruder, process materials which contain considerable levels of gluten (semolina flour, or hard winter wheat flour, vital wheat gluten, etc.) develop the great elasticity and gas holding properties which are characteristic of hydrated and worked glutenous doughs.

To strengthen that dough (make it chewier) one could increase the level of gluten, could extrude at higher moisture contents, and perhaps at higher total protein content, or modify the pH on the acid side.

If the end product desired was a dense, but fully cooked

breading, for example, the basic raw material selected might be a hard winter wheat flour (or second clear), wheat middlings, wheat bran, etc.

To weaken that extrudate (make it more tender) it would be practical to replace the hard winter wheat flour with a wheat flour of lower gluten strength and lower protein content, like soft spring wheat flour. The use of reasonable levels of lipids or emulsifying agents will also tend to weaken the dough. Oat flours, being high in fat content, also tend to weaken the dough strength. Extruding such ingredients or mixtures at low moisture content in the dough, coupled with higher temperatures of extrusion, will also combine to weaken the dough strength of the final extrudate.

If a light and frothy extrudate is desired (crispy), it would be useful to select cereal ingredients with low gluten strength and low fat content. True, fat weakens a dough but it also provides plasticity and hardness, neither of which is desired in a crisp and tender end product. Ingredients, then, which meet the dual requirement for low fat content and low gluten strength would include degerminated corn and/or rice flour and/or wet milled corn starch. Such products would be extrusion cooked at low moisture contents and relatively high extrusion temperatures for maximum crispness.

Modification of pH of ingredients can also have a toughening effect, if such modification is on the acid side. Acid modified starches, for example, provide fluidity in a dough, lower its viscosity and cause an extrudate to slow its thermoplastic tendency toward hardening as it cools. Conversely, pH modification of starches on the basic side sometimes will increase the viscosity of the extrudate and may have a tenderizing effect on starches.

Selection of ingredients to provide desired nutritional characteristics in the final product represents one of the greatest responsibilities and opportunities for the producer of a new "engineered" food (6). However, I want to emphasize that the formulator of a newly engineered food needs to satisfy both the structural and textural properties, as well as the nutritional requirements of the new food he is engineering.

CONCLUSION

HT/ST extrusion cooked foods are being developed from a wide range of raw materials (cereal flours or granules, starches, tubers, legumes, fibers, proteins, fats, minerals and vitamins) to produce an ever widening variety of fabricated foods. Emulsifiers, flavors, surface active agents, pH modifiers, glutens, food colors and gums are all used as tools of the food scientist, to produce end products which have desired nutritional, functional and organoleptic characteristics.

Extrusion cookers are designed with great flexibility to create food products which are needed today and which may be wanted tomorrow. Protein enrichment of foods is being offered today by many manufacturers of breakfast cereals and to a lesser extent by manufacturers of snacks, dry soup mixes, cookies and wafers.

HT/ST extrusion cookers pasteurize and irreversibly bond process materials together, and are designed to make the cleaning of processing equipment easy, quick, and simple. Sanitation of processing equipment can combine with pasteurizing processed materials through extrusion cooking to produce the great shelf life of extrusion cooked foods.

HT/ST extrusion cookers are thermodynamically efficient, and can process engineered foods at lower processing costs per ton than any other method of thermal processing.

Tools are at hand in the design of extrusion cooking system; in the selection of processing controls, and in the selection of ingredients and microingredients to produce a wide range of economical fabricated foods. Equipment and labor costs per ton of production are much lower on HT/ST extrusion cookers than on the older methods of thermal processing.

REFERENCES

1. Slesser, M. -- V. Sci. Fd. Agric., 24 1193-1207 (1973).
2. Berry, S. -- Congressional Record, 92nd U.S. Congress, S. 2430 (February 23, 1972).
3. Scrimshaw, N.E. in "Protein Enriched Cereal Foods for World Needs" -- Edit. M. Milner, p. vii, The American Association of Cereal Chemists, St. Paul, 1969.
4. Johnston, G. L. in "Technical and Practical Processing Conditions with Single Screw Cooking Extruder", 1978 -- International Seminar; Cooking and Extruding Techniques -- ZDS -- Solingen -- Grafrath, Germany.
5. Smith, O. B. in "World Soybean Research" Edit. Hill, L. D.; "Textured Vegetable Proteins"; Interstate Printers and Publishers, Inc., Danville, Illinois, p. 861-882 (1976).
6. Smith, O. B. in "New Protein Foods, Vol. 2" -- Editor A. Altschul, Acedemic Press, Inc. (New York and London).

Index

F

G

H

I

L